THE GRAND SCUTTLE

The Sinking of the German Fleet at Scapa Flow in 1919

DAN VAN DER VAT

BIRLINN

In memory of my father

This edition published in 2016 by
Birlinn Limited
West Newington House
Newington Road
Edinburgh
EH9 1QS
www.birlinn.co.uk

First published in 1982 by
Hodder and Stoughton Limited, London

Reprinted 2021

ISBN: 978 1 84341 069 0

British Library Cataloguing-in-Publication Data
A catalogue record for this book is available from the British Library

Typeset in Bembo MT by Hewer Text UK Ltd, Edinburgh
Printed and bound by Clays Ltd, Elcograf S.p.A.

Contents

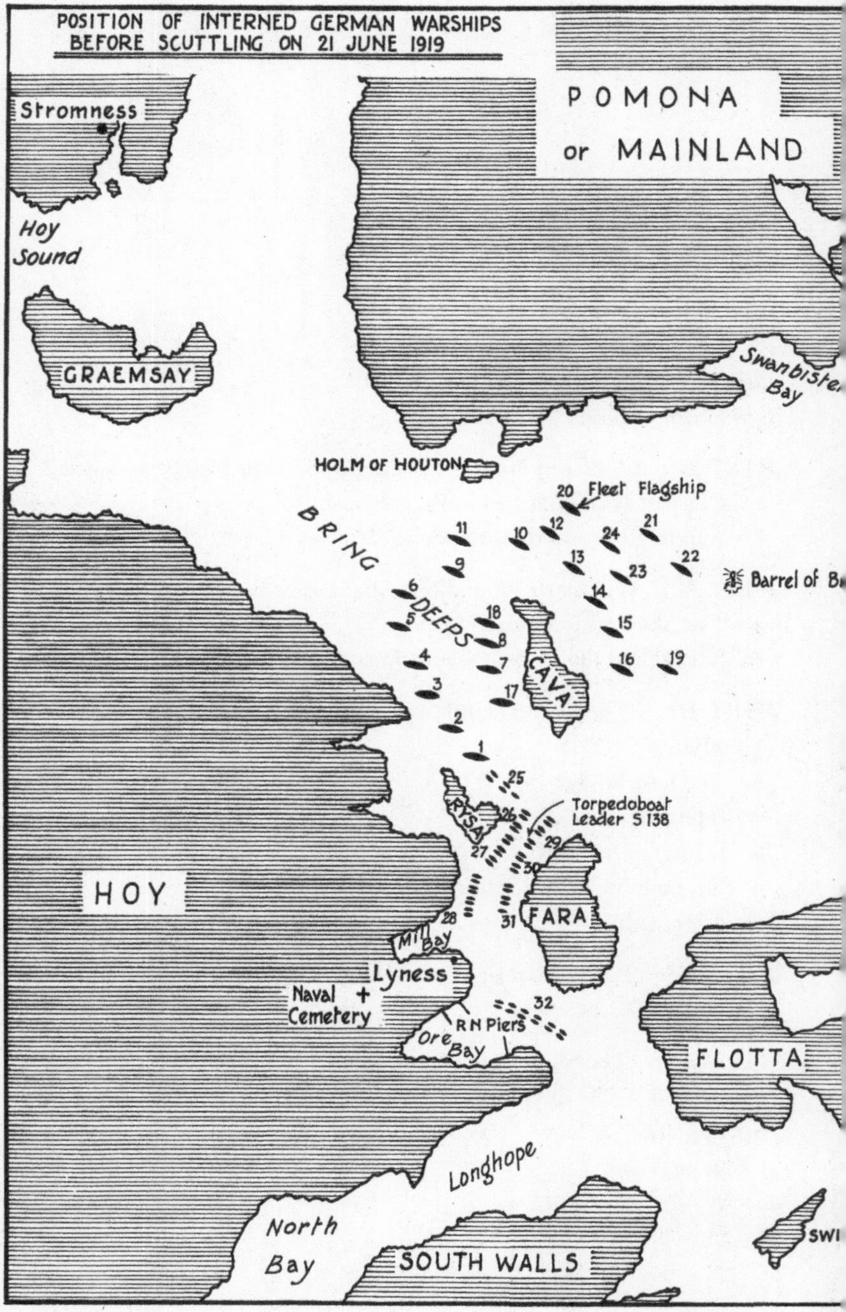
POSITION OF INTERNED GERMAN WARSHIPS
BEFORE SCUTTLING ON 21 JUNE 1919
POMONA
or MAINLAND
Stromness
Hoy
Sound
GRAEMSAY
Swanbister
Bay
HOLM OF HOUTON
20 Fleet Flagship
BRING
DEEPS
11
10
12
24
21
9
13
23
22
Barrel of B
6
14
5
18
8
15
4
7
16
19
CAVA
3
17
2
1
25
FYSA
26
Torpedoboat
Leader S 138
27
29
30
HOY
28
31
FARA
Mill
Bay
Lyness
Naval
Cemetery
32
R N Piers
Ore
Bay
FLOTTA
Longhope
North
Bay
SOUTH WALLS
SWI

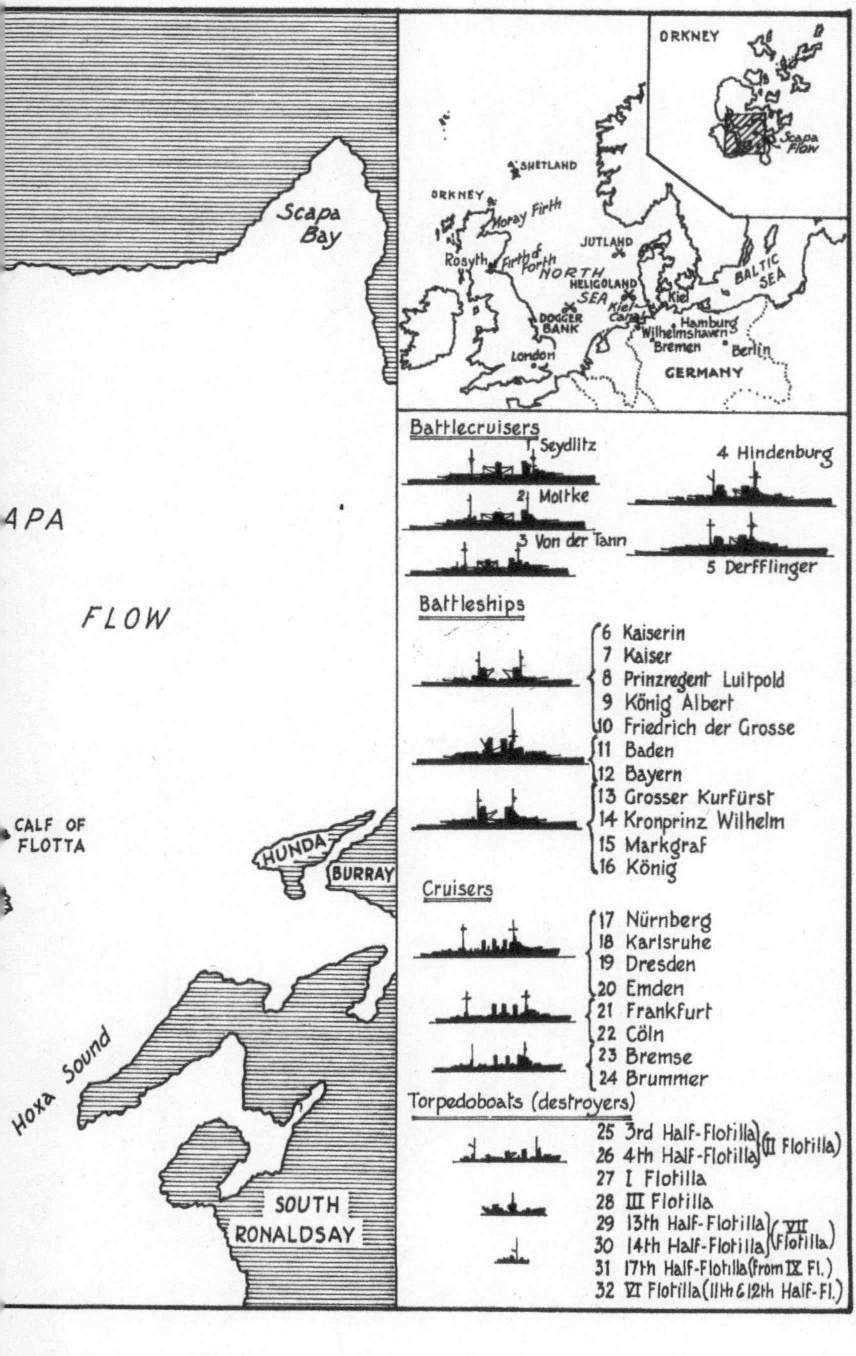
Scapa Bay
APA
FLOW
CALF OF FLOTTA
HUNDA
BURRAY
Hoxa Sound
SOUTH RONALDSAY
ORKNEY
Scapa Flow
SHETLAND
ORKNEY
Moray Firth
Rosyth
Firth of Forth
NORTH SEA
JUTLAND
HELIGOLAND
DOGGER BANK
Kiel
Kiel Canal
BALTIC SEA
Hamburg
Wilhelmshaven
Bremen
Berlin
London
GERMANY
Battlecruisers
1 Seydlitz
2 Moltke
3 Von der Tann
4 Hindenburg
5 Derfflinger
Battleships
6 Kaiserin
7 Kaiser
8 Prinzregent Luitpold
9 König Albert
10 Friedrich der Grosse
11 Baden
12 Bayern
13 Grosser Kurfürst
14 Kronprinz Wilhelm
15 Markgraf
16 König
Cruisers
17 Nürnberg
18 Karlsruhe
19 Dresden
20 Emden
21 Frankfurt
22 Cöln
23 Bremse
24 Brummer
Torpedoboats (destroyers)
25 3rd Half-Flotilla
26 4th Half-Flotilla
(II Flotilla)
27 I Flotilla
28 III Flotilla
29 13th Half-Flotilla
30 14th Half-Flotilla
(VII Flotilla)
31 17th Half-Flotilla (from IX Fl.)
32 VI Flotilla (11th & 12th Half-Fl.)

Preface

FOR THREE YEARS before this book was first published in 1982 I lived with the sonorous names of dead ships. The names are royal, military, commemorative and honorific or merely functional. *Seiner Majestät Schiff Seydlitz* eventually became my favourite because she should have sunk at the Battle of Jutland in 1916, but refused, because she bore the burden of leading the German High Seas Fleet into internment at the end of the First World War, and because she resisted more than forty attempts to raise her from the bottom for scrap, after the pride of the Imperial Navy scuttled itself at Scapa Flow in Orkney on 21 June 1919.

On that day there occurred the greatest single loss of shipping since Man first sat astride a log and floated away from land. The Persian fleet met its end at Salamis, and the Pacific Fleet of the United States was smashed at Pearl Harbor, but both those disasters resulted from enemy action. The Spanish Armada was scattered and destroyed, but the main reason for its doom was a providential storm. The German High Seas Fleet survived a cataclysmic war almost unscathed, but the bulk of its strength and tonnage was destroyed by order of the German Admiral in command at the time. The scale of the loss, over 400,000 tons of the finest warships then in existence, seventy-four vessels of which fifty-two actually went to the bottom under the eyes of the enemy, is unique in itself. That it was an act of *self*-destruction based on a misapprehension only compounds the uniqueness of a story which is extraordinary by any standard. It is also surprising that the story of this fleet has never before been told in logical form, from its inception through its construction and frustration in peace and war to its humiliation, its self-immolation in the name of honour, its salvage and its still continuing value as a source of uncontaminated steel: from the idle speculations of an under-occupied naval staff officer called Tirpitz to Man's first ventures into space. That is the theme of this book.

The reader is entitled to know why and how the author, a most un-naval person, came to attempt to fill this gap. I am a reporter by profession. *The Times* saw fit in October 1978 to send me to Orkney to cover a seal-cull which never happened. This incidentally had the effect of enhancing my long-standing claim to be one of the most successful if coincidental protectors of the British seal population. The two previous culls I had been sent to cover by another newspaper years before at the Farne Islands off Northumberland were also cancelled. If the non-event in Orkney was anything at all, and it was not much, it was what is known as a 'colour story'. There was plenty of that: white seal pups with big brown eyes like mermaidenly labradors on grey rocks in the grey-green sea, and also Greenpeace, the ecological pressure group to which the credit for the eventual cancellation mainly belongs. Ready to report at all times, but with the best will in the world somewhat short of events to write about, I found myself with time on my hands in a remarkably beautiful part of the world and I took the chance to explore, so rarely available on out-of-town assignments. Half of Fleet Street seemed to have signed the visitors' book at the splendid Cathedral of St Magnus in Kirkwall, the Orkney capital. It also paid to venture further afield.

Thanks to Churchill Causeway, you can drive to the southern extremity of the islands and look down upon John o'Groats from the north, seven miles distant, which must in a small way be comparable with looking at the Arctic icecap from underneath. The causeway was built by Italian prisoners in the Second World War.

As you drive along it you see blockships, sunk to protect the Scapa Flow anchorage, ringed by the Orkney islands, against submarine attack. Further explorations on the western side of Scapa Flow by boat and on land reinforce the recollection that this great protected area of water was the main shelter of the British fleet in both world wars. It is a small step from this elementary rediscovery to look for detailed relics of Scapa Flow's brief but crucial role as the marine 'garage' of the British fleet. It is an even smaller step to learn that the German fleet was 'parked' here at the end of the First World War, and that this is where it sank rather than surrender.

I was forcefully reminded of this when I went to the stone town of Stromness with its canyon-like streets, hidden behind the port which is still its main livelihood. It was only natural to visit the tiny Stromness Museum and admire the magnificent obsession of the bygone taxidermist of local origin whose work occupied most of the space. He stuffed everything from seagulls to seals and was only narrowly forestalled from tackling a whale by local fishermen whose needs were more immediate. In summer 1974 the little museum found room for a temporary display of pictures and relics of the German ships sunk in Scapa Flow in 1919. The special exhibit proved so popular that it became permanent, and this is what I saw. I was intrigued by the pamphlet on sale which told the story of the Imperial Navy in terms of the amazing salvage operation which followed its destruction. I was sufficiently stirred to visit the Orkney central library in Kirkwall a day or two later to ask what books had been written about the stupendous naval suicide of 1919.

The library staff (to whom I am indebted) led me to their inner sanctum, which contains an enormous collection of books wholly or partly about Orkney that may be consulted only on the spot, and produced a respectable pile of books which mentioned the scuttle, in passing or at some length. These works described everything from the Anglo-German arms race in ships which preceded (and largely caused) the First World War to the salvage of the German fleet after it. They were all admirable in their different ways and I have drawn heavily upon them; but my visit to the Orkney Library and subsequent researches in London (the bug had taken hold) did not reveal an account of the wasted fleet from construction to destruction and beyond. By this time *The Times* had temporarily closed down, and I was looking for something to do until it resurfaced, which I soon concluded would take a long time. I resolved to prepare a synopsis of a book about the sunken fleet in the hope of attracting the publisher's advance which was essential to finance the necessary research in Germany and Orkney. Thanks to Graham Watson of Curtis Brown Ltd, my literary agent, since retired, a publisher was eventually found, and my accidental interest in the affairs of the Imperial Navy was translated into a commitment.

As it turned out, I had just enough time to complete the research before *The Times* resumed publication. The economic recession, however, hit the publishing trade hard, and the project foundered when the original publisher scuttled, abandoning the production of general interest non-fiction. The more than half-completed book seemed to be sunk. Then my flatteringly enthusiastic new literary agent, Dinah Wiener, persuaded Ion Trewin of Hodder and Stoughton to undertake its salvage – my very best thanks to him.

The main focus of the research was naturally the West German Federal Military Archive in the beautiful city of Freiburg. I am most grateful to the staff of that institution for their help in tracing whole mountain ranges of largely unexplored and unused material which forms the basis of Part III, the core of the book. The files relating directly to the lost fleet would, if stacked, have made a column of paper two metres high; the files which touched upon it in passing might have filled the coal-bunkers of SMS *Seydlitz*, and I was able to do no more than taste them. The German material is raw and extremely comprehensive; the files in the British Public Record Office at Kew in Greater London, whose staff were no less helpful, were neat, rather thin and suspiciously sanitised. But then it is a German fleet of which I write, and it was only to be expected that the original owners of the ships would have infinitely more information about them than the former enemy, for whom their destruction at German hands under their noses was in the end no more than an embarrassing incident. The sparseness of the British files attests to the official British attitude about the events in Scapa Flow in 1919: least said, soonest mended.

I am indebted to a number of German naval veterans for their help. They include Commander Yorck von Reuter, son of the Commanding Admiral; Vice-Admiral Friedrich Ruge, former Head of the Federal German navy and Scapa Flow internee, who gave me permission to make use of his memoir; and Seaman Werner Braunsberger.

I should like to say a special word about Herr Braunsberger, who was located for me by the very helpful Deutscher Marinebund (German Naval Federation, the veterans' organisation in Wilhelmshaven). Herr Braunsberger, living in retirement in Bielefeld, wrote to me in response to a notice placed for me in the Marinebund's magazine, saying he had a personal, written account of his experiences aboard the interned battleship *Kaiser*. I wrote back at once, explaining why I was anxious to see it. Herr Braunsberger wrote again saying that after due reflection, he felt unable to entrust his proud memories 'to a non-German'. I was able to explain that I was of Dutch origin and birth and that I intended to try to be as neutral in my account as the Dutch had been during his war. I hoped he would agree that justice could be done to the subject only if I had full access to first-hand information. The reply was a large envelope containing his account, loaned to me 'with a heavy heart and after much internal conflict'. I was most touched and felt honoured; I hope more earnestly than I can say that I have kept my side of the bargain, both in making use of his invaluable material and in trying to produce a balanced and fair account of the scuttle, as I promised the dignified and polite old gentleman in Bielefeld I would try to do.

I received much help from Orcadians, too: their names are mentioned in the text that follows. I should also like to thank Gerald Meyer, editor of the remarkable weekly *The Orcadian*, who left Balham, South London, to spend a few days in Orkney and is now well into his fourth decade there. He gave me access to his files, publicised my project and put me in touch with people who had information about the events of 1919.

One result of his assistance came in the form of a letter from an exiled Orcadian living in Aberdeen, who had an interesting theory about the Cockney slang word 'scarper', meaning to escape or run off. 'Knowing the tendency of the southern English to put in an "r" where it should not be (to the fury of the Scots!), I vouchsafe the suggestion that "scarper" is derived from Scapa because of the scuttling of the German fleet.' It would be delightful if this were true; unfortunately the word derives from the Italian *scappare*, meaning to escape, and came into English early in the last century. But it was a nice thought. I am no less grateful for the help, encouragement, interest and advice I received from family, friends and colleagues in Germany and in Britain.

So this is the story of the *Seydlitz* and her sisters, with iron names such as *Derfflinger, Von der Tann, Friedrich der Grosse, Karlsruhe, Nürnberg, Emden* . . . or just numbers – *G40, V129, S32, B109, H145.* They were built as instruments of destruction and death, but there was beauty in them too, and their fate is an unabashedly romantic story. This book contains no fiction or 'faction' and all quotations are documented. I have done my best to tell it as it was. If there are errors, they are mine alone.

INTRODUCTION

Scapa Flow, 21 June 1919

MIDSUMMER'S DAY 1919 brought to the northerly latitudes of the Orkney Islands the kind of fine weather that inspires artists, opening as if to infinity a panorama usually closed in by cloud, fog, haze or rain. The dawn was very, early, slow and hardly perceptible because at this time of year, if the sky is clear, the islands almost become an honorary province of the land of the midnight sun. In the great anchorage of Scapa Flow, shielded by the islands, the sea was calm and exceptionally blue, reflecting an open sky.

As people began to go about their daily chores on land, those with a view of the Flow once again saw, in the finest possible light, a spectacle which had hardly changed in seven months. The best view was from the hillcrests of Hoy, the only mountainous island of the group. Immediately to the east of it, drawn up in neat rows at its feet, lay seventy-four grey warships – five battlecruisers, eleven battleships, eight cruisers, fifty destroyers. They were German, and they had been there since shortly after the Armistice ended the First World War in November 1918. They had not been surrendered; they were in internment, floating hostages for German compliance with the Armistice, which was about to be translated into the Treaty of Versailles by the fraught, last-minute negotiations still continuing on that day. The ships remained German property, manned by German skeleton crews, and constituted the main strength of the Imperial Navy's High Seas Fleet, undefeated in battle. They were under the surveillance of the Royal Navy, but there were no British guards aboard; under international law they would have been trespassers.

Few Orcadians will have given the display more than a passing glance. They were accustomed to seeing great assemblages of warships in the Flow, because it had been the principal anchorage of the British Grand Fleet throughout the war so recently ended. The enemy warships had long since lost their novelty. The only way you could easily distinguish them from Royal Navy vessels at a distance was by their lack of a flag at the stern. The Imperial Naval ensign, with its black cross and eagle, was forbidden in British waters. Only command flags flew from the masts. Closer to, the great floating fortresses and the slim destroyers showed clear signs of advanced neglect, the rust and the spinach-coloured marine vegetation drooling from their anchor-chains attesting to the demoralisation and decay of internment. Further over, British white ensigns could be seen on a squadron of five battleships, the main component of the guard-force, which showed signs of getting up steam.

A handful of tiny patrol-boats and tenders chugged slowly round the familiar rows of anchored ships. A few German sailors could be seen moving about on the decks in the glorious weather so seldom provided by the capricious climate of Orkney. The normally bleak and desolate view was transformed by the sun into a scene of exceptional beauty. Even the German sailors, who had been denied access to shore for more than half a year, acknowledged that. A few of them saluted Midsummer by putting on white uniforms, a rare sight indeed in Scapa Flow. Apart from the special clemency of the weather, it looked like business as usual as those ashore and those afloat turned their minds to breakfast.

Below the untidy decks of the German ships, however, there was a new tension, something in the air hitherto unknown in all the dreary days of confinement. Over the past three or four days, odd things had been happening. Officers, engineers and senior petty officers had been unusually active, clambering into the less accessible parts of the ships and emerging begrimed. They had been seen going round the ships with heavy tools, opening every door they came to and pinning it back, loosening hatches, opening portholes and breaking rods which connected valves below the waterline to control levers above deck. Hammering and sawing was heard coming from below, echoing through the sparsely manned ships. Officers had taken to sleeping aboard destroyers, unoccupied for months and moored alongside their sister-ships; and they took not only food, drink and blankets, but also a sledgehammer or a huge wrench. Only a few of the 1,800 or so men in the fleet knew officially what was going on, and they had been ordered to keep it to themselves. Most of those in the know were captains, and key officers and senior ratings whose co-operation was essential.

The German naval mutiny in the dying days of the war had destroyed the trust between officers and men and exacerbated the difficulties of imprisonment afloat beyond measure. The officers had experienced humiliation and hardship in bringing the ships to their fateful rendezvous with the Grand Fleet off the Firth of Forth in November 1918 prior to internment, and the unrest had continued on most of the ships thereafter. But the more alert crews and individual ratings either guessed or demanded to know what all this unusual activity meant. Aboard some ships, everybody knew; on others, nobody but a small handful. Those who did find out what was going on experienced in many cases a resurgence of the old spirit which had enabled them to fight so well when they had the chance, before the stalemate in the North Sea in the latter part of the war damaged their morale. These revitalised men seethed with an excitement they had not known since going into battle. They could not know, however, although many had guessed, that the British had drawn up detailed plans for the simultaneous seizure of every ship the moment the Armistice ended. As more and more men learned or deduced the secret behind all the activity, the enemy's intentions, whatever they might be, lost much of their relevance. The uncertainty of indefinite incarceration – the depression caused by enforced inactivity – were lifting at last. The Armistice, as everybody knew despite being starved of news from home and forced to rely in the main on four-day-old copies of *The Times* doled out in miserly quantities to the command by the British, was about to end at last, after a long series of extensions.

Yet it could just as likely be followed by a renewal of hostilities as by peace. The conditions imposed on a chaotic and prostrate Germany were extremely harsh, and such leaders as the country had were finding them hard to swallow. The Allies were insisting on the surrender of all the interned ships. On this there was absolutely no chance of a compromise. Those internees who were interested enough to follow events at Versailles as best they could, and there were many, knew that the career of the High Seas Fleet must now come to an end in this hated anchorage in British waters. It could not fight because its guns had been spiked; it could not run away because its engines were run down. It would not be allowed to go home. Only two possibilities remained: that they would be handed over to the enemy, or that they would not, slipping away from his grasp in the moment of consummation of the Allied victory. 'The flag officer in charge, German ships at Scapa', as the British referred to him, Rear-Admiral Ludwig von Reuter, expected hostilities to resume at noon when he awoke on the morning of 21 June. The last newspaper he had seen, dated 16 June, reported an Allied ultimatum to this effect after the German government had refused to sign the Treaty. He had already made his dispositions and his choice. Nothing happened that morning to make him change his mind.

After breakfast on that fine and fateful Saturday morning, lines of children began to form on the quayside at Stromness, the little port in the north-western corner of the Flow. They were waiting to file aboard the Admiralty tender *Flying Kestrel* which had tied up alongside. Eventually some 400 children of all ages from the local school were being taken for the special day out planned by their teachers and sanctioned by the Royal Navy – a trip round the interned fleet. The crew and the teachers were relieved and the children delighted by the splendid weather, which promised a calm passage and a chance for a really good look at everything.

Soon after 9 a.m. all was ready, the last stragglers counted and safely brought aboard, and at about 9.30 a.m. the tender put to sea. She steamed slowly southwards through the lines of moored ships, on a course roughly parallel with the eastern coast of Hoy. The big ships came first, towering over the squat water-tender. Later came the rows of destroyers in the channel between Hoy and Fara. The acting excursion steamer passed Lyness, the ugly little port where the principal Royal Navy shore installations spoiled the scenery. The children chattered and pointed, waving at the German sailors they saw looking down at them, who responded as they saw fit. After passing through the last isolated line of German destroyers at the end of the channel, the *Flying Kestrel* put about for the return voyage. It was noon. So far the school outing had been merely exhilarating.

Half an hour after the *Flying Kestrel* had begun its trip southwards, the British battle squadron with its escort of cruisers and destroyers sailed westwards into the open sea for routine torpedo exercises. The only British warships left in Scapa Flow now were a couple of destroyers to back up the handful of small craft on patrol among the interned enemy fleet.

At this point – 10 a.m. – Admiral Reuter was seen to appear on the quarterdeck of his flagship, the cruiser *Emden*. The acutely observant watcher would have been struck by his full-dress uniform. One of his officers approached, saluted and addressed him for a minute or two. The Admiral paced up and down, pausing frequently to study various ships through a telescope. At 10.30 a.m. he turned to a sailor waiting nearby with the large notepad of a signaller. Shortly afterwards, a string of flags appeared over the ship. They said in plain language to the initiated: *Paragraph 11. Bestätigen.* Paragraph eleven. Confirm. How strange that the Germans should be sending a message like that outside permitted signalling hours. But the Admiral's telescope was now busier than before.

Bernard Gribble was an artist who specialised in marine scenes. He was inspired by the wonderful weather that morning to beg a ride on one of the patrolling naval trawlers so that he could sketch the German ships. Suddenly he noticed that boats were being lowered, against British standing orders, and told an officer. The naval lieutenant paused, then cried: 'I've got it! I believe they're scuttling their ships!'

It was noon. Sixteen minutes later, the battleship *Friedrich der Grosse* turned turtle and sank.

At that moment the children on the *Flying Kestrel* felt the engines beneath their feet shaking the vessel as she piled on all the steam she could muster. It was their first intimation of the drama that was about to unfold as the tender raced for Stromness to put them ashore – a history lesson in the making which none of them would ever forget. The kaleidoscope of destruction stunned their eyes; but in old age it was the accompanying cacophony of sound made by a dying fleet that these children of long ago so often recalled first. The outward trip offered them a close-up view of a long familiar picture, a huge still-life of helpless, hopeless hulks. The return journey tore the scene to shreds before their eyes and ears like a massive earthquake.

These 400 children witnessed the greatest single act of destruction at sea ever known, the result of a decision by the only man ever to set out to sink a navy at a stroke. The German High Seas Fleet came to a fitting end. If we now go back to its beginnings, it will become clear why this is so.

PART I

A Study in Folly – The Anglo-German Naval Arms Race

Place a military force as strong as possible . . . in the hands of the King of Prussia; then he will be able to conduct the policy you want. It is not conducted by speeches and shooting-matches and songs. It is conducted only with blood and iron.

– Prince Otto von Bismarck, speech to the Prussian legislature, January 1886

It is on the navy under the Providence of God that the safety, honour and welfare of this realm do chiefly attend.

– King Charles II, Preamble to the Articles of War

CHAPTER ONE

German Naval Expansion Ends British Isolation

THE ONCE MAGNIFICENT warships which died in Scapa Flow that Midsummer's afternoon in June 1919 had been the main strength of what was, vessel for vessel, the most modern and powerful navy in the world. They were also the embodiment of one of the greatest geopolitical and strategic errors in history and the product of an obsession sustained for two decades in defiance of reality. Nor is it the comfortable hindsight of several generations later that produces such a judgement. The mistake was seen clearly at the time and was eventually acknowledged, though much too late, by Admiral Tirpitz, one of the two men most responsible (the other was his sovereign, Kaiser Wilhelm II). That this protracted and devastatingly costly blunder was also one of the main causes of the First World War makes it one of the most terrible aberrations on record.

The phenomenal expansion of the German navy from 1898 to 1914 was both a product and one of the principal expressions of the power of the new Germany. Otto von Bismarck had united the various German states and statelets under Prussia, whose military pre-eminence in Europe made it the irresistible 'locomotive' of German unification. After three short and successful wars against Denmark, Austria and France, which brought extra territory to the Reich, the first Kaiser, Wilhelm I, was crowned with lavish insensitivity in the Hall of Mirrors at the Palace of Versailles in January 1871. In the period of consolidation that followed, Germany, already the greatest military power in Europe, built up a strong economy on the basis of a new and efficient infrastructure, financed to a significant extent by French war reparations. Wilhelm I died in 1888 at the age of ninety and was succeeded by Crown Prince Friedrich, who was terminally ill and ruled for just ninety-nine days. Had he lived, the course of history would, almost certainly, have differed radically because he was a convinced liberal and a committed opponent of militarism and Bismarck's 'blood and iron' policy. As it was, Wilhelm II, who could hardly have been less like his father, became Kaiser at the age of twenty-nine. He soon grew tired of the restraining hand of the old 'Iron Chancellor', who was as wise as he was ruthless, and pushed Bismarck into resigning in 1890. The death of Friedrich and the departure of Bismarck was a double disaster for Europe and was seen as such at the time.

In his later years as Chancellor, the post created for him in 1871, Bismarck had devoted himself to the maintenance of the status quo he had done so much to bring about. With the free use of duplicity and the mailed fist, he also made Germany a stabilising force in Europe, as most of its neighbours recognised much of the time. As soon as he retired, Europe began to divide into two camps. The secret German 'reinsurance treaty' with Russia was allowed to lapse in 1890; in 1891 Russia concluded an entente with France, bringing that rapidly reviving country out of the isolation to which it had been consigned by Bismarck's policy since its defeat by the Prussians in the war of 1870. In 1892, the entente was extended by an agreement on mutual military assistance, and a second power-block arose in Europe alongside the Triple Alliance of Germany, Austria and Italy. Britain, as usual, kept up her 'splendid isolation', which had suited Bismarck but irritated Wilhelm II. Under the Kaiser, Germany became the ever-demanding cuckoo in the European nest, driving the British to swallow their distaste for Continental entanglements by entering into the 'Entente Cordiale' with France in 1904 and coming to terms with Russia in 1907, thereby creating the 'Triple Entente'. The British and the Germans made a number of attempts to reach an understanding – and almost succeeded. In the end the growth of the German navy destroyed the possibility.

In 1894 an obscure staff officer of the Naval Supreme Command, Captain Alfred von Tirpitz, with some help from colleagues, wrote a memorandum to his superiors arguing the case for a strong fleet. The captain, a technical expert on torpedoes, had already grown grey in Germany's then very junior service, which he had joined in 1865, and he had the knack of persuasive argument of a case on paper. Tirpitz (1849–1930) started work on his thesis when he was Chief of Staff to the Admiral in command of the Baltic in 1891. The following year he was posted to the Supreme Naval Command in Berlin, where he completed his work on it. The document came to the attention of Wilhelm II and fired the imagination of the erratic ruler as few things did before or after. Such was the moment of conception of what was to become the German High Seas Fleet. The main thrust of the Tirpitz paper was that Germany was now a world power of the first rank with worldwide colonial and commercial interests and therefore needed a world-class navy capable of fighting decisive battles at sea. Germany needed a battle fleet.

Bismarck had shown no interest in naval matters. Prussia, and subsequently the German Empire it dominated, vested its security and its strength as a Continental power in the world's most modern and powerful standing army. This was a logical, geographical imperative as simple and inescapable as Britain's reliance upon the world's largest navy. And Prussia had been able to forge the German Reich and even acquire a respectable scattering of colonies round the world without needing anything more than a token navy, a point conveniently overlooked by the navy lobby when it became a power in the land at the end of the century.

But now policy seesawed. In 1865, the Prussian Parliament rejected a ten-year plan for enlarging the fleet and making Prussia a second-class naval power. Two years later, the decision was reversed and a ten-year programme for the construction of sixteen armoured ships, a mixture of large and small vessels, was adopted. The scheme was abandoned in 1870 because of the war with France, which produced just one small naval incident, a skirmish between a German gunboat and a French despatch boat off Cuba. The ten-year plan was revived in 1872, adopted in 1873 and completed, with some reductions, in 1883. It enabled Germany to show the flag on the high seas, watch her overseas interests and have the potential to make hit-and-run raids in the event of war with a major European power.

Count von Caprivi – a General be it noted, not an Admiral – was put in charge of the Admiralty in 1883 and initiated a four-year programme for building light cruisers. These were designed for raids on enemy commerce, a sound strategy for a land power such as Germany, as the First World War was to show. Caprivi also proposed a large force of torpedoboats and recognised the need for a 'High Seas Fleet' to support the cruisers in distant waters. It was during his term of office, which ended in 1888, that work began on the strategic Kiel Canal linking the Baltic and the North Sea across Schleswig-Holstein. Although Caprivi was very much on the right lines in his thinking about Germany's limited naval needs, the Reichstag showed little enough interest even in those; the General-turned-Admiral himself had no enthusiasm for battleships in the belief that their days were numbered because of the torpedo. On that he was only half right: it was not until the torpedo was combined with the submarine that the invincibility of the battleship ended, ironically, just as the battleship itself became the grand obsession of the Admirals and statesmen of the principal maritime powers. German naval policy and strategy remained ambivalent during the period from 1890 to 1897, when Admiral Hollman served as State Secretary in the Imperial Navy Office. A poor orator, he was unable to make a favourable impression in the

Reichstag. But in 1897 he was succeeded by Tirpitz, now an Admiral, as the Kaiser's chosen head of the navy.

Wilhelm II had had no strong thoughts about his navy until he saw the Tirpitz memorandum in 1894, though he had always loved ships. When he was Crown Prince, he had tried his hand briefly at designing battleships. He saw the Great Naval Review marking Queen Victoria's Golden Jubilee off Spithead in 1887 and was most impressed. He had also read a seminal work of the time, *The Influence of Seapower upon History* by the American Admiral Mahan, published in 1890. This was also the year that Heligoland, the small island of vital strategic importance in naval terms which guards Germany's North Sea coasts, was ceded to the Reich by Britain in exchange for colonial concessions in East Africa. The Kaiser envied Britain's overwhelming naval supremacy but was not originally hostile to it because he did not regard Britain as a rival, still less a potential enemy.

The personality of the German Emperor was a curious and unattractive one, even though he had, when he chose, a certain charm. As the heir of the Prussian ruling house, the Hohenzollerns, he was steeped in military tradition and much of his young manhood was spent in military training and in the company of dashing young officers. The Imperial Court had a highly military flavour and was distinguished by its grand uniforms. Wilhelm II was emotional and arrogant, romantic and obsessive, erratic, a poor judge of men and impossible to work with except for the yes-men he tended to gather round himself. By and large he was a pompous ass who had no clear idea of what he wanted for the mighty Empire misfortune had chosen him to rule. He was short on both intellect and common sense and showed a great lack of tact.

The system of Imperial government he inherited was not designed to cushion the workings of his personality defects; rather it tended to magnify them. Even though the Prussian landowners, the Junkers, did not exercise the monopoly of influence they had exerted in pre-Imperial Prussia, Prussia dominated the Reich and so did its system of government, essentially a feudal oligarchy. The key to the system was the Chancellor, who ruled alone under the Kaiser. There was no Cabinet. This was all very well when the Chancellor, who was also Prime Minister of Prussia, was Bismarck and left by the Emperor to get on with the job. But when lesser men were chosen and Wilhelm II proved unable to delegate (not that he wished to), the system could not cope and was bound in the end to break down. A contemporary German socialist, Stampfer, described his country under Wilhelm II as the best run but worst ruled in Europe. The Emperor, never slow to change his mind, listened to the Junkers, the Generals and the leaders of the economy he gathered round him at court rather than to the head of his Government. Under him Germany became unpredictable because he was unpredictable. There was at least one thread of consistency through his reign, however: Germany was always demanding. At home, the political system was primitive and the class system archaic; yet the economy was not constricted by these factors and flourished with strong Government aid and encouragement, while social welfare was exemplary if paternalistic and the general standard of living rose dramatically.

To fulfil their scheme for a German battle fleet, the Kaiser and Tirpitz had to start from scratch. Tirpitz had followed up his memorandum with another, shorter paper in 1895 (at the Kaiser's request) which further expounded his case for a German High Seas Fleet. In this he set out his 'risk theory': Germany should be strong enough on the high seas to be capable of inflicting serious damage on the world's most powerful fleet, even in a losing battle. This strength would deter the leading naval power because its own fleet would be so reduced in such an encounter that its world maritime supremacy would be lost and the basis of its power destroyed. The argument smacks of today's nuclear deterrent theory. It would therefore avoid the risk of attacking the German fleet, which would thus become a powerful means of extracting concessions or of forcing an alliance (which to the Germans meant a master-and-servant relationship). Although she was not named, all this of course was aimed at Britian: no other interpretation was possible.

German relations with Britain, however, first deteriorated sharply when Wilhelm II saw fit to interfere in the rising tension between Britain and the two Boer republics in South Africa. The two empires had already been at odds in 1892 over the building of the Baghdad Railway across Ottoman Turkish territory, but the Kaiser's impulsive telegram of effusive congratulation to President Kruger of the Transvaal in January 1896 on his resolute repulse and crushing of the Jameson Raid roused public opinion on both sides of the North Sea for the first time. The Kaiser actually wanted to send troops to help the Boers but was happily dissuaded, not least because he had no means of safely delivering them, a point which he regarded as a strong argument in favour of expanding his navy. As far as he was concerned, Germany must change its policy from internal to external expansion, and to do that it had to acquire a navy other powers would be obliged to take seriously, so that Germany could move from the European to the world stage.

When Tirpitz took over the Admiralty in June 1897 he set out to implement his ideas. Nine months later, the first German Fleet Law was passed (in March 1898) after a two-day debate in the Reichstag which ended with a vote of 212 in favour and 139 against. At that point, Britain had twenty-nine modern battleships with another twelve under construction compared with Germany's thirteen and five respectively. Now Germany was committed to the construction of seven further battleships and two large cruisers. Allowing for the decommissioning of superannuated ships, this would give Germany nineteen battleships, twelve heavy cruisers, thirty light cruisers and eight armed coastal defence vessels all by 1903, at the end of the five-year period covered by the Fleet Law.

Germany's pre-1898 battleships had, on average, been a third smaller than Britain's, being all under 10,000 tons with less than half the coal-bunker space. But in November 1897, the month in which the draft of the Fleet Law was published, Germany seized the Chinese harbour of Kiaochow as a naval base in the Far East, a place of little value for its existing navy which could only become significant after major naval expansion. The seizure was part of the general plundering of China fashionable at the time.

Tirpitz was not only an unusually effective speaker for a professional naval man, he was also something of a propagandist and did not rest content with having the ear of his Emperor on naval matters. He also needed to mobilise public opinion in support of a big navy. Thus the Navy League was set up in 1898, supported financially mainly by the Krupp steel and armament company of Essen in the Ruhr, the district with the most to gain from a large naval programme. The League was a runaway success, drawing in nearly a quarter of a million members in three years. Wilhelm II was its patron and those in search of power and influence soon took up the cause. To justify its existence and its argument for naval expansion, it generated as much anti-British feeling as it could, a task made easier by the Anglo-Boer War. The Kaiser often let it be known how different his policy would have been had he had a navy capable of intervening. As it was, Germany was obliged to remain neutral.

The Boer War unleashed anti-British sentiment in most of Europe, not merely in Germany. Liberal feeling had advanced far enough to see the (white) Afrikaners as the underdogs and hapless victims of British Imperialism and its lust for gold (but not far enough to worry about the blacks in South Africa: the European right to rule non-white peoples was still a long way from being questioned). But Wilhelm's attempt to create a Continental league to oppose British expansionism outside Europe came to nothing. In Germany, however, Anglophobia was particularly exacerbated by the runaway success of the Navy League's propagandising, which was particularly effective in the schools. Unable to intervene, the Kaiser fumed on the sidelines and saw how the British, with their domineering navy, could defy world opinion by moving troops and supplies at will and sealing off their stubborn enemy by blockade. Things would be different in twenty years, he told his Foreign Minister, Bülow: by that time Germany would have a navy nobody, not even the British, could ignore.

The British, meanwhile, preoccupied as they were with their embarrassing flounderings against the Boers and unable to end the game quickly even after stacking the deck in their own favour, did not reciprocate the German hostility; rather they found it puzzling and wondered what they had done to deserve it. Distrust of Germany was slow to grow and made significant advances only when reinforced by fear. Yet the first German Fleet Law was introduced just as the British used the threat of their naval strength to stop the French acquiring control of the Upper Nile Valley (what was known as the Fashoda incident). The lesson was not lost on the Germans.

The British unwittingly played into the hands of the naval lobby once again just as Tirpitz drafted his second Fleet Law at the beginning of 1900. The Royal Navy, acting on a false alarm, boarded a German mail steamer on its way to South Africa. The British thought the ship was carrying volunteers and war material to Delagoa Bay to help the Boers. An international row laced with stiff diplomatic Notes ensued, with strong public opinion exhibited on both sides. The British eventually recognised that they had been in the wrong and paid compensation.

Tirpitz was emboldened to make direct and open reference to the British and their naval dominance in introducing this second Fleet Law, something he had avoided in the first, even if the message could be read plainly enough between the lines. The object of the second Bill was to double the strength of the German navy in sixteen years by constructing three ships a year. By 1920 the High Seas Fleet would consist of two flagships, four squadrons of eight battleships each, eight heavy and twenty-four light cruisers. Tirpitz's stated goal was to create a navy capable of fighting a battle in the North Sea against the British. Bülow, speaking in the Reichstag in support, said Germany had to have a large navy to be able to deal with Britain, and the German fleet would be developed with an eye to British policy.

This was a fateful decision for Germany. The first Fleet Law alone would have made Germany a force to be reckoned with on the high seas; the second brought her into open competition with Britain in a way the British could not afford to ignore. It was the point of no return. Germany's shipbuilding capacity was much smaller than Britain's, so that new dockyards had to be developed for the naval programme. The enormous investment required of Government and shipbuilders was such that, once the programme began, deceleration became very difficult and cancellation inconceivable for economic reasons. Thus the fatal flaw of inflexibility was introduced, and with it the inevitability of confrontation with the increasingly anxious British. But the specially built modern dockyards gave the Germans the advantage in ship design capacity: when the British started counter-building to maintain their superiority, they had to work within the limitations of their older yards. The race to build dreadnoughts, the new and larger generation of capital ships, accentuated the difference. As it is much easier to lengthen a dock than to broaden it, the British produced narrower ships than the Germans, whose vessels enjoyed the extra stability and strength conferred by their distinctly broader beam. Starting almost from scratch as they were, the Germans could produce better-designed ships in purpose-built yards incorporating all the latest equipment and techniques.

The ships provided for in the first Fleet Law were larger than anything the Germans had built before; those planned under the second were larger still and would double the number of battleships. A higher maintenance budget was also envisaged. Tirpitz insisted that the yards had to have a guarantee of continuous production, and only a Law could provide it. The envisaged building programme could be adhered to or overtaken by an even larger one; but it could not be reversed once the Bill became a Law. While the Germans failed to foresee the consequences abroad, particularly in Britain, of their decision, the British never understood why the Germans behaved as if their naval legislation was as immutable as the Ten Commandments and incapable of moderation in the interest of international understanding. There was no room here for the British art of compromise. A Law imposed a duty.

The most vocal opposition to the naval programme came from the Social Democratic Party (SPD), whose leaders objected on grounds of cost and, with more accuracy than they could have known at the time, of the risk of war. But their anti-navy rallies were far surpassed by the powerful lobbying of the Navy League. But to appease the opposition, the Government conceded five large and five small cruisers as well as a reduction in the proposed cruiser reserve, and the Law went through. It was now Germany's duty to see it observed, and her spokesmen thereafter quite genuinely explained the programme as the inescapable fulfilment of a legal obligation. Only the Germans could see it that way – indeed, *would* see it that way. The scope for misunderstanding was infinite, and the British and the Germans alike, for reasons of national character and national interest, were now doomed to talk at one another in a dialogue of the deaf.

British incomprehension of German legalism was matched by German underestimation of the British instinct to dominate the seas. Like many another British institution, the Royal Navy at the turn of the century was the result of spontaneous growth. There was no such thing as a long-term naval construction programme.

Britain's decisions on building ships were reactions to what other naval powers were doing. Underlying this approach was the philosophy of the 'two-power standard', whereby the Royal Navy must always be at least equal in strength at sea to the second and third most powerful navies in the world combined. The strategy behind this was both simple and fundamental to Britain's survival as an imperial and trading nation. The overwhelming geographical advantage of their island conferred upon the British immunity against invasion, provided only that they maintained a strong navy. If that navy matched or surpassed the combined navies of any possible alliance of other major powers against her, Britain could afford to maintain only a small standing army, to pursue her policy of staying out of Continental entanglements adopted after the Napoleonic Wars, to develop and protect her leading position in world trade, and to police a worldwide empire without fear of interference. The benefits of naval supremacy were so enormous and so vital to national interests that Britain was bound to react to a new challenge in that area, even if she started slowly and tried at first to make a new naval expansion unnecessary by diplomacy.

The version of logic deployed by the Germans in favour of a large navy was so full of non-sequiturs that it is essential to look at it more closely to try to establish exactly what Germany wanted. There was at least one great flaw in Tirpitz's stated aim of building a navy which could give at least as good as it got; ship for ship, a navy large enough to deter the leading naval power from taking it on for fear of losing its mastery of the seas. He had to concede (he could hardly avoid doing so) that Germany must pass through a 'danger zone' before she had assembled a High Seas Fleet of capital ships numerous enough to constitute such a deterrent. This period would last from the adoption of the building programme to its completion. The gamble was worthwhile, the naval lobby argued, because it would have the effect in the end of converting Germany from a virtually land-locked Continental power into a world power of the first rank. Germany, they said, should have a navy commensurate with her enhanced stature. It was her right as a world power and an essential part of her dignity. The trick, therefore, was to persuade the German people that the Reich could establish itself at sea without actually matching the strength of the Royal Navy.

In support of this questionable thesis, it was argued that Germany *needed* a powerful fleet, even though she had rubbed along without one very comfortably hitherto, because she needed peace, protection against possible blockade and threats to her trade, and a means to defend her new overseas colonies and seaborne commerce. The new navy need not be as large as the British because the Royal Navy had too many worldwide commitments to be able to concentrate enough ships to be sure of defeating Germany without losing supremacy. An enlarged German navy would make it impossible for Britain to rely on being able to take on any conceivable naval coalition, before or (especially) after a maritime confrontation with Germany. Britain might well defeat the High Seas Fleet, but she would then be helpless against an alliance of, say, France and Russia. The High Seas Fleet was therefore to be seen as a powerful lever against Britain.

The deterrent theory was all very well as far as it went, except that at the end of the nineteenth century the British hardly needed to be deterred. Tirpitz himself even used the apparent lethargy of the British as a point in favour of accepting the risks of the passage through the danger zone, a fine piece of logic-chopping.

During the Boer War the Germans successfully exerted disproportionate pressure on Britain to obtain a colonial interest in the Pacific island territory of Samoa, previously shared by the British and the Americans. (They did so, once again, without a world-class navy, be it noted.) The British were puzzled by the vehemence with which Germany pursued its 'claim' in Samoa, which they saw as a case of using a sledgehammer to crack a nut; they could not see what the Germans stood to gain.

The danger zone could in fact be said to exist already as Germany was in any event helpless against the Royal Navy. But what if the British woke up and reacted by counter-building? The only naval building policy they had ever followed was to keep an eye on what other naval powers were doing and then to build enough new ships to stay ahead. Surely the danger zone would then be extended indefinitely.

Further, the risk theory held good only if Britain maintained her traditional isolation and distaste for Continental entanglements. The Germans assumed that Britain and Russia would never see eye to eye because of their rivalry in the Far East, especially over the buffer-zone between their two empires (the 'Great Game'). Britain was unlikely to come to terms with France because of the Franco-Russian Entente. The idea that the British might be *driven* to change their policy of isolation for the very reason that their naval supremacy faced an unprecedented threat was either ignored or discounted by the Germans.

The greatest threat the Germans identified was that the British might send another Nelson to 'Copenhagen' the nascent High Seas Fleet in what we would now call a pre-emptive strike. (Nelson knocked out the Danish fleet in 1801 to prevent it joining up with Napoleon's navy, and the verb 'to Copenhagen' was coined to describe such preventive strokes. German naval propagandists used this incident as an awful warning of Germany's helplessness without a large navy; at the same time Tirpitz dismissed the danger as remote because the British were supine. A decade before the First World War, however, powerful voices in Britain would call for a Copenhagen against Germany, including Sir John Fisher who, as First Sea Lord, built up and modernised the Royal Navy with its first long-term construction programme.)

The German reading of the mood of the British was that they would do too little too late. Insofar as there was to be no new Copenhagen, they were right; insofar as Britain reacted slowly and reluctantly, at first, to the German naval programme, they were right; where they were wrong was in their habitual underestimation of opponents in general and of the British in particular, especially when the latter became aroused, however late in the day, by a threat to national survival. They also undervalued the huge industrial resources available to British shipbuilders, who were already in a position to enable Britain to stay ahead of the expensive naval game that was about to begin for as long as it lasted, given only a national will to go on ruling the waves.

Whereas Tirpitz had started his naval programme without overt reference to the British and mentioned them openly only in the context of the second Fleet Law in 1900, even then with some circumspection, other influential Germans were less inhibited and talked of the real possibility of war with Britain. The naval challenge was the greatest threat to the balance of power cherished by Britain in almost a century, and would be a major contribution to its destruction. The British meanwhile were somewhat jealous (though not as much as the Germans thought) of German industrial and commercial growth, which was faster than their own, and worried about general Anglophobia in Europe caused by the Boer War. The Germans were simply envious of Britain's world role, of which her navy was the mainstay. Some Germans, however, were already aware that their major Continental neighbours had greater incentives to oppose Germany in the longer term than to tolerate her. The French wanted Alsace-Lorraine back after losing it in the War of 1870; the Russians were competing with Germany's ally, Austria, for influence in the unstable Balkans. Now the Germans were setting out openly to reduce Britain's naval supremacy, a threat so vital that, once recognised, nothing would prevent the British from countering it, even if it meant forming alliances (as they had done before, after all). And they were conceding in advance that they could never actually catch up. But the Kaiser resolutely ignored the warnings of his more perceptive envoys and ministers, and backed the stubborn Tirpitz to the limit.

The Admiral, meanwhile, took it as read that Germany would remain pre-eminent in military strength on the Continent. He put his faith in the Royal Navy's inability to concentrate in the North Sea because of its conflicting role of policing a worldwide empire. Even if the British counter-built, the Royal Navy would surely soon run into manpower problems. On this basis it was not impossible to imagine that the High Seas Fleet might even muster a *greater* strength in the North Sea long enough to make all the difference. Only a change of heart by the British over entanglements with Continental powers could upset this calculation. That is precisely what happened. Yet as late as 1904, the year of the Entente Cordiale between Britain and France, Tirpitz was saying that the High Seas Fleet would outnumber the Royal Navy's available strength in the North Sea by 1912 (and would also outgun the entire French fleet).

But it was not only the British who finally failed to accept the German assumption that all the other pieces on the board would remain meekly in place while the High Seas Fleet grew. France still wanted her lost provinces back; Russia was still at odds with Austria in the Balkans and worried about Germany's interest in the Near East and its dealings with the Ottoman Empire. The open naval challenge to Britain was bound to make both the other major European powers review their calculations about relations with Germany in the light of the principle that the potential enemy's new enemy could become a friend. The Germans' misplaced confidence that they could somehow stalemate the eternal superiority of the Royal Navy was based on the assumption of permanent dispersal of British warships round the world; but it was open to the British to offset this in three ways. They could at the very least strengthen the fleet in home waters by reducing the policing of the Empire to a minimum (and encouraging the willing Dominions to share that burden); they could build more ships; and they could call up massive reinforcements from the Mediterranean at short notice – especially if appropriate arrangements could be made with other powers, notably France, but also Italy, the ambivalent (being basically Anglophile) member of the Triple Alliance with Germany and Austria. For the Germans to suppose that they could pick off the Royal Navy piecemeal was to suppose that the British would take no notice of a rise in Anglo-German tension until it was too late or would do nothing in response to a general deterioration in the international atmosphere in the direction of war.

The British in fact were friendly towards Germany in 1898, the year of the first Fleet Law, and Joseph Chamberlain suggested an Anglo-German alliance, which was rejected. The Germans bided their time because they thought such factors as Anglo-Russian rivalry in the Far East would sooner rather than later make the British return as suppliants for an understanding with Germany, which would mean better terms. Both countries took a cynical view of each other's fidelity to treaties anyway (not without reason). The Germans had noted the misleading tenor of a sensational, unsigned article in *The Saturday Review* (in September 1897) which argued the inevitability of an Anglo-German trade war. Tirpitz referred to this isolated piece of journalistic kite-flying for years afterwards; it was the kind of speculation that was commonplace in German publications but rare in British. The Germans consistently overestimated British jealousy of their increased strength in industry and trade. German competition in fact had a salutary and healthy effect which enabled the British to be sure of holding their own by the turn of the century. Meanwhile Germany by this time was second only to India as a market for British goods, the loss of which would have been a serious setback. The trade potential of the German colonies was minimal and no ground for war. There was clearly room for both countries to expand their trade without getting in each other's way. The British had much more reason to worry about the industrial growth of the United States, but managed to lose no sleep over that either.

In the early days of the growth of the German navy, therefore, the British took a relaxed view. They showed understanding of Germany's desire to be in a position to guard her merchant marine and colonies. Britain probably realised rather earlier than Germany and its navy lobby that it was in a position to prolong the danger zone indefinitely. In 1900 German trade rivalry was contained; in 1901 there was talk in Britain of a leisurely nature about easing tension with Russia and consequently France; in 1902 there was moderate unease in Britain about Germany's naval ambitions, especially when the Kaiser compared Germany's future naval strength with Britain's for the edification of the Reichstag. In 1903 the British took their first concrete counter-measure by choosing a new North Sea naval base at Rosyth. This was a momentous step, even if the main motive for it appeared to be the pressing need for more port space for the Royal Navy rather than countering Germany. Hitherto, the British fleet had been concentrated upon the south-east of England in terms of bases, on the traditional assumption that the main potential enemy was France. The range and speed of contemporary warships made a base on the eastern coast of Scotland useful against any potential enemy but, inescapably, particularly so against Germany – which in September of the same year assembled the first of its two planned home-based fleets, to be ready at minimum notice for action in or from home waters. Meanwhile the British were moving closer to the French in the contacts which were to lead to an Entente Cordiale in 1904, and the explosion of the risk theory.

By this time, the 'two-power standard' was rather more than a century old without anyone in Britain being sure of what it meant, beyond a feeling that the Royal Navy should have more than France and Russia combined. Even the Admiralty was unclear about what this meant in tonnage and types of warship. The Naval Defence Act of 1889 instituted a five-year building programme of replacement, with the need for policing the Empire by cruisers in mind. In 1898 the Government still held to the view that the Royal Navy should be a match in numbers of ships, and preferably superior in fire-power, to any combination of two other navies; enough in fact to deter *three* other powers taken together. In the same year the Royal Navy formed a new 'flying squadron' for the Channel. Even before Germany began to rock the boat by expanding her navy, the British noted that the Americans, the Italians and the Japanese were also developing their own naval capacity to such an extent that a three-power standard might impose itself, in defiance of the practicable. In 1900 concern about German plans was voiced in Parliament for the first time. In that year, without significant decline in the value of money over the intervening period, the naval estimates rose to £26m compared with £15m in 1890. In 1904 they reached £37m. German naval expenditure rose from £5.7m in 1897 to £16.5m in 1908 and £20.1m in 1909, with a far larger proportion going on construction than on maintenance. From 1904 British calculations were dominated by the need to pay special attention to the rise of the German navy. By the end of 1905, the Royal Navy was stronger than ever, and it was possible to think in terms of a standard in capital ships of two to one, plus 10 per cent to take account of the need to police the Empire. But the standard remained undefined.

While the British over the turn of the century distrusted Germany, they showed no appetite for a trial of strength. By contrast, the Boer War had fuelled Anglophobia in Germany to an unprecedented degree, and it was expressed in the strongest terms as akin to a national issue. The Kaiser wanted nothing less from Britain than that she should become the fourth member of the Triple Alliance and abandon her distaste for such involvements. The navy lobby took full advantage of the burgeoning divergence of the two countries, although their interests still managed to converge from time to time.

The British and the Germans worked well together in defence of the principle of imperialism. Both co-operated with other nations in the suppression of the Boxer Rebellion in China in 1901 and in a classic piece of gunboat diplomacy against Venezuela in the following year for failing to pay its debts and maltreating foreign traders. (On both occasions the British felt the Germans used rather more than the minimum necessary force.)

Meanwhile the two countries pursued their distinct foreign policies, the new factor being cautious British readiness for formal arrangements, however limited, with other powers. The Triple Alliance of Germany, Austria and Italy was still in being at the turn of the century, but was more of a liability to Germany than an asset. The arthritic dual monarchy of Austria-Hungary was dangerously weak and unmistakably on the decline. And although Italy had renewed the Alliance for a second time in 1897 (the treaty was concluded in 1882 and had been renewed in 1887), she openly declared that she would not respond to a call from her partners for help against Britain or even against France, in defiance of German wishes and the text of the agreement. In 1902 Britain concluded a three-year treaty with the rising sun of Japan. This required each side to show 'benevolent neutrality' to the other in the event of the latter coming under attack by a third party, and to intervene in favour of the other if it were attacked by a combination of powers. There was a risk here that Britain could be drawn into war with France if Russia attacked Japan (which, given their rivalry in the Far East, was quite likely, as would soon be shown), and then called on the French as allies to help them. In such an event, Japan was entitled to ask for British help. In 1905 the Anglo-Japanese treaty was simplified to provide for active support from one party to the other if it were attacked by just one other power. Thus to avoid the risk of conflict with France, the British turned their minds to an understanding with the French. The result was the Entente Cordiale, which fell a long way short of an alliance, but rapidly produced a workable understanding.

London and Paris came to terms on their respective spheres of influence outside Europe, especially in North Africa. Europe was omitted, nor was there any undertaking to provide support in the event of war or any provision for military or naval co-operation. Nonetheless, it was a seminal event in European history. The ceremonial side of this cautious and limited agreement to be friends, in which King Edward VII played such a leading role, was far from empty. The 'hands across the Channel' gesture of the two ancient rivals was soundly based on an over-riding common interest: mistrust of Germany. It was not to be long before the General Staffs of the two countries began informal conversations about what they would do in the event of a Continental land war.

For the British, it became realistic to think with confidence of rapidly reinforcing their new Channel Fleet (created in 1902 by combining the previous Channel and Home Fleets) by virtually denuding the Mediterranean of British ships, leaving the area to the French in exchange for protecting their Atlantic coast. This was to give the lie to Tirpitz's risk theory, as well as being a historic realignment of forces in Europe. As early as the end of that momentous year of 1904 the British Admiralty ordered a permanent strengthening of the Channel Fleet by calling up ships from distant stations and reducing the Atlantic Fleet based on Gibraltar. The French stepped up expenditure on their own navy at the same time. The Kaiser resolutely ignored repeated messages from his envoys in London that the British were motivated by fears for their naval supremacy and not at all by economic considerations. The German resolve to build up the navy was stiffened, to British surprise.

CHAPTER TWO

Acceleration and the Prelude to War

As the British and the French began to move towards an understanding, the Japanese started a war with Russia by a surprise torpedo attack on the Russian fleet in Port Arthur on the Chinese coast, which they then blockaded from February to August 1904 until their troops reached the port overland. As a result the Russians with their six battleships broke out of Port Arthur, fought an inconclusive action with Admiral Togo's four battleships and returned to port. The stalemate that followed did not last for long. In December the Russian squadron was completely destroyed by a combination of Japanese army artillery and the patiently waiting Togo.

In the meantime the Russians had despatched their Baltic fleet, headed by four new battleships, on an enormous voyage round Europe, Africa and Asia to help their beleaguered Pacific squadron. The ill-trained and badly run fleet from Kronshtadt entered the English Channel on the night of 21–22 October 1904 and mistook a flotilla of Hull trawlers for a hostile navy. The Russians opened fire, hitting five fishing vessels. The Channel Fleet was ordered out and cleared for action. War seemed imminent between Japan's ally, Britain, and the ally of Britain's new-found friend, France, but the danger passed as the Russian ships straggled southwards, incompetently fouling the nets of another fishing fleet on the way past Dover. Small wonder then that the Russians were annihilated by Togo in the momentous Battle of Tsushima on 27 May 1905, and ceased to exist as a naval power. Japan won the war and Russia had a foretaste of the 1917 Revolution. The decisive battle was won by superior discipline, speed and above all heavy gunnery. Other world navies set about learning the lessons.

The Germans tried to exploit the Anglo-Russian Dogger Bank incident by suggesting a joint action against Britain. The Kaiser met the Tsar in July 1905 and proposed a treaty of mutual aid in the event of either being attacked in Europe. Later the Tsar wrote to the Kaiser to say that the treaty could not apply in the event of a French attack on Germany because France was Russia's ally. The Kaiser dropped the subject and, in disgust, concluded that the Russians, for what they were worth, were now in the Anglo-French camp. After Russia's disaster with the Japanese the Tsar was in no mood for a new adventure of whatever kind. In naval matters Britain was by no means idle. In the month of the Dogger Bank incident, Admiral Sir John (later Lord) Fisher was appointed First Sea Lord and started work on Britain's first ever integrated naval expansion programme. He soon earned a rebuke from the King for suggesting a Copenhagen against Germany. When in 1905 others in authority made the same suggestion, there was panic in North Germany followed by a wave of renewed support for a strong navy which was exploited by the Navy League and Tirpitz.

The Kaiser deliberately set out to test the Anglo-French entente early in 1905 by making a series of intentionally incomprehensible demands for a German say in the affairs of Tangier which, as an enclave in Morocco, was within the French sphere of interest. The British, however, loyally supported the French in the contrived dispute, to the surprise and disappointment of the Germans. The Kaiser's arrival in Tangier in March started a crisis over what was really nothing more or less than a German attempt to bully the French. Germany demanded a conference about the future of Morocco, which duly took place at the beginning of 1906 in Algeçiras. The Germans were isolated. The neutral Americans and even Germany's nominal ally, Italy (which had a secret treaty with

France from 1900, recognising French interests in Morocco in exchange for concessions over Tripolitania), sided with France, which thus emerged with a handsome moral victory instead of the humiliation planned by Germany. Ham-handedness had once again cost the Kaiser dear. But Morocco rankled to such an extent that Germany almost triggered off a war over it six years later, with the Agadir incident.

Anglo-German naval rivalry intensified and went into its critical phase from the end of 1904. Edward VII had seen his cousin Wilhelm's new fleet at Kiel and was duly impressed. The British people were now thoroughly aroused about German naval plans and felt as strongly about them as the Germans felt about the British.

Skilfully riding the tide, Tirpitz drew up a supplementary Fleet Law revising the second Fleet Law of 1900. Six heavy cruisers and forty-eight destroyers were to be added to the 1900 programme, and the battleships included in that programme and not yet built were to be increased in size (and in cost). A small sum was allocated for research into submarines. Tirpitz warned that another supplementary Law might be needed in a few years. This one went through in November 1905.

Within three weeks the British First Lord of the Admiralty (navy minister), Lord Cawdor, laid down Admiralty policy in a memorandum. The British let it be known that the pace of their building programme would be related to Germany's and would be planned one year at a time. Four capital ships would in any event be completed annually but provision was made for stepping up production in case of need and plans were drawn up for accelerating the deployment of naval reserves in war. The Admiralty also set up a committee on Design.

The Anglo-German naval arms race had begun in earnest.

The most important technical development had already taken place at Portsmouth on 2 October 1905. On that day work began on laying the keel of a battleship to be named HMS *Dreadnought*, the apotheosis of the Nelsonian ship of the line, whose ascendancy was rendered unexpectedly and, in view of the expense and effort involved, disappointingly short by the rapid development of the submarine and the aeroplane.

The significance of the *Dreadnought* was seen immediately it was launched, evidence of how closely naval developments were followed at the time and how fashionable navies were. The Paris edition of the *New York Herald*, the paper for Americans in Europe, reported in its edition of 11 February 1906:

> PORTSMOUTH, England – The battleship *Dreadnought*, the last word of progress in naval architecture, was launched here yesterday in a ceremony presided over by King Edward. The effect it is likely to have on the future shipbuilding of the world's navies must indeed be considered to inaugurate a new era. Already, before complete details of its design, construction and equipment are known, other nations are preparing to duplicate the type, which has been brought about by the lessons of war in the Far East. The designers of the ship have combined the largest possible number of heavy guns with a displacement consistent with the existent facilities for docking such a vessel.

The battleship had grown rapidly, not only in size, and erratically over the preceding half century as the quality of steel, armour-plating, engine-power, design and above all armament developed alongside the general technological advances of the nineteenth century. Some very strange ships took to the water, as British, French, Italian and American ship designers were given their heads and as one naval power copied and adapted the work of others. The British and the Americans came to the conclusion at about the same time that the 'all big-gun' ship was the naval weapon of the future, and the United States started work on two. But it was the relentless drive of the First Sea Lord 'Jacky' Fisher which saw to it that the British got there first in an unprecedented construction effort. The *Dreadnought* was fully fitted out and ready for sea-trials one year and one day after the first plates were laid. This was an astounding achievement, reflecting awesome self-confidence on the part of the British, as well as industrial prowess, organisational ability, technical leadership and a readiness to experiment with new ideas.

The new ship, with a displacement of just under 18,000 tons, was less than 10 per cent larger than the last of the pre-dreadnoughts, as older battleships soon became known, and at £1.75m was less than 20 per cent costlier. Her main armament was ten of the latest 12-inch guns, each capable of firing 850lb shells a distance of 18,500 yards. Three of the five turrets were positioned on the centre-line with one on either side amidships. This meant that the *Dreadnought* could fire a broadside of eight shells on either side and six ahead (but only two astern: His Majesty's ships did not require massive firepower astern as they were unaccustomed to flight, it was haughtily explained at the time). The heaviest armament previously carried by a battleship was four 12-inch guns of which only two could fire ahead. Thus one *Dreadnought* was worth two older battleships at long range and three when firing ahead. Her secondary armament was very secondary indeed, consisting of light guns for use against destroyers (these proved too light and were replaced on later ships of this type by heavier weapons). She was designed round her guns as a floating artillery battery.

Nor was the revolution she represented confined to the enormous firepower she carried. Her 11-inch belt of armour, made from the latest German steel bought from Krupp, could withstand both severe pounding and the enormous shock of eight huge guns firing simultaneously. This aspect too was a new departure: the latest advances in rangefinding, sighting and fire-control were built in, so that all eight guns in a broadside could eventually be laid, aimed and fired from one point by one gunnery officer positioned in his 'control top' observation post halfway up the tripod-based foremast.

Hitherto, as at the Battle of Tsushima, guns had been fired separately without centralised control and their enormous range far outstripped earlier rangefinding capacity. The 12-inch gun had been preferred over the 10-inch for the *Dreadnought* because its higher velocity shells had a flatter trajectory, conferring greater accuracy at long range (the optimum being 10,000 yards or a staggering five nautical miles). The bold decision to fit turbine engines, first tried out in destroyers only seven years earlier, gave the new ship much greater power for less weight and less cost. Her maximum speed was twenty-one knots, a match for the best cruisers of the day. Conventional reciprocating engines would have cost £100,000 more and added 1,000 tons to displacement. She also looked different with her uncluttered lines, and her faults – not enough heavy armour below the waterline, too light secondary armament and the low-slung pair of side-turrets which could not be used simultaneously in a broadside – were soon ironed out in her descendants.

The first purpose-built, all big-gun ship did not, however, lack critics, inside and outside the navy. The new type rendered half a generation's investment in earlier battleships obsolete at a stroke, some of them only recently commissioned. Other nations would surely follow the lead and Britain's traditional overwhelming superiority must, therefore, be eroded as a result. Others wisely drew attention to the already growing twin menaces of the torpedo and the submarine which could make such an enormous concentration of effort and investment in the new monsters foolhardy. Traditionalists disapproved of the implicit change in strategy involved in choosing speed before strength, over-specialisation of armament before versatility, long-range engagement before short range.

Gone were the days when the British waited for others to take the initiative and relied on industrial superiority to 'go one better, faster'. Fisher and his supporters took the view that other powers were bound to opt for the quantum jump in firepower represented by the *Dreadnought* (as the Americans were already doing) precisely in the hope of rendering Britain's earlier types of capital ship obsolete and reducing her lead. Britain could not afford to let others open up a gap when it came to exploitation of a super-weapon which all the progressive experts knew was bound to come soon as the result of the latest technological developments. For all her temporary pre-eminence, the *Dreadnought* was directly descended from her predecessors. Her design was a breakthrough because it overcame so many of the drawbacks of earlier capital ships at the same time, thanks to a fortuitous as well as fortunate convergence of progress and resolution. The decision to go ahead was an expression of British determination to retain command of the sea whatever the cost. There were to be no half-measures either.

Four months after work began on the battleship, the keel of a heavy cruiser variant, to be named HMS *Invincible*, was laid – the first of three. These soon became known as battlecruisers. They too were given 12-inch guns (eight instead of ten) under centralised fire-control (only guns of the same type could be co-ordinated in this way) and a distinct margin of extra speed (twenty-five knots). They were slimmer, forty feet longer, and carried armour little more than half as thick as that of the battleship version, all to suit their role as protectors of merchant ships, as raiders, as reconnaissance ships, as reinforcements for the line of battleships, able to raise firepower at short notice, and as leaders of the chase. The great guns were essential, which meant that armour had to be sacrificed for speed, a risk which, in general, proved worthwhile. They were classed as capital ships and were soon sharing the new generic appellation 'dreadnought' in the ever-changing league tables of world navies.

The Germans, who had launched fourteen battleships since 1900, were temporarily thrown into disarray by the birth of the dreadnought. They had hoped that the election of a Liberal Government in place of the Conservatives at the start of 1906 might take the edge off British determination to stay ahead in the naval competition, especially since the new government had ambitious social reform plans which needed funding. The Liberals were indeed interested in arms control, but not at any price. Whatever their misgivings, they adopted a policy of building four dreadnoughts a year for the time being. And it was the Liberal Government which, in the light of the Moroccan affair, allowed the general staff to start consultations with the French and Belgian army commands in 1906. The relationship with France was rapidly becoming a *de facto* alliance, and the new Foreign Secretary, Sir Edward Grey, would not allow himself to be diverted from it for the sake of possible disarmament and the ever more elusive chance of an accommodation with the difficult Germans. Committed though they were to spending cuts, the Liberals consistently supported the Royal Navy's supremacy, though not without misgivings.

The Germans, brought up short by the dreadnought development, laid down no new capital ships for a year. Then in November 1907, on the basis of another amendment to the Fleet Law, they announced that they would build three a year to Britain's four. The British were thoroughly alarmed, having failed earlier in the year at The Hague Peace Conference to get anything out of the Germans on disarmament or at least a reduction in the tempo of the naval arms race. In 1908 the Liberals asked again for a cut in tempo and suggested a ratio of three to two in capital ships (something of a retreat from the two-power standard but nothing like enough for the Kaiser and Tirpitz). The British had the temporary consolation of seeing the Germans floundering to catch up: their first two dreadnoughts, the *Nassau* and the *Westfalen*, were inferior imitations, although the Germans were in the end to surpass British designs; and the new breed of capital ship was too large for the strategic link and short cut between the Baltic and North Sea bases, the Kaiser Wilhelm (or Kiel) Canal, which had to be enlarged at great effort and expense. The size and number of large new ships turned out by the great modern dockyards soon outstripped the limited capacity of Germany's small number of naval bases to accommodate them. They had to be expanded and deepened and new barracks and other shore facilities had to be provided. The British had no such disadvantages and also had a much larger building capacity. But the economically sound British practice of using existing yards to expand the Royal Navy had the one great drawback of imposing limits on design, particularly on the breadth of beam of the largest ships, as we have seen above. The German fresh start gave them the advantage here.

The first four German dreadnoughts were slower than the British ones. They had twelve 11-inch guns, but the turrets were so positioned that only eight could be used for either broadside. But their armour was distinctly superior. They also had a much smaller range (dispelling any last illusion as to their real role: deployment in the North Sea), which conferred structural advantages: smaller coal-bunkers meant that more and stronger bulkheads could be incorporated.

Another amendment to the Fleet Law in 1908 called for the laying down of four dreadnoughts a year until 1911 and then two a year. Heavy cruisers already budgeted for, but as yet unbuilt, were to be battlecruisers of the new type. Germany was now committed to a fleet of thirty-eight battleships and twenty battlecruisers of this class. The British, even Fisher, were not unduly concerned at first: the Royal Navy was four times the size of the German and had in 1908 a five to two lead in dreadnoughts at sea or on the stocks. At the end of 1908, however, the Admiralty's complacency was punctured by accurate reports that the Germans were also increasing their capacity to make the guns needed for the new ships, which individually took twice as long to manufacture as the ships themselves. By expanding gun-building capacity and stockpiling other components, as was happening at a new plant installed by Krupp, Germany was in a position to step up steeply the rate of ship production and appeared to be ready to do so despite assurances and denials. The Admiralty concluded that the Germans could pass the British in dreadnought strength as early as 1912 and called for a doubled programme of eight starts in 1909. The government approved four with provision for another four if the Germans forced the pace. General British distrust and dislike of Germany came close to panic, and the jingoistic cry was raised, 'We want eight and we won't wait,' with the Opposition Conservatives in the van. The German naval estimates for 1909 were 22 per cent up on those of 1908, their biggest jump ever, a fact which fuelled what was now the dominant British national obsession.

Germany had one of her own: the fear of isolation and the ancient nightmare of the war on two fronts. Having failed to separate Russia from France in 1905, and Britain from France in 1905–6, the Germans drove the Russians in 1907 into favourable consideration of a rapprochement with Britain. The Germans antagonised Russia with their Baghdad Railway project, a line from the Mediterranean to the Persian Gulf via the Valley of the Euphrates. This would have benefited Turkey, with whom Russia was at odds over the Slav states in the Balkans (as she was with Germany's ally, Austria). Britain refused to join the railway consortium, which would have resulted in a threat to British dominance of the Gulf. But the British were at odds with the Russians over a clash of interests in Persia, and there remained the traditional rivalry over the buffer zone between Russia and India. Russian feelers had, however, been put out towards Britain as early as autumn 1905; in August 1907 the two came to terms on their spheres of interest, not only in Persia but also in Tibet and Afghanistan. Europe was not mentioned. Although they did not approve of the terms at first, the British thereby became a member of a Triple Entente with France and Russia.

The rapprochement was a major blow and strengthened German resolve to build a navy strong enough to break out of this ring; the Navy League reached a membership of 900,000. The Kaiser's disingenuous denial that Germany wanted to challenge British naval supremacy, in a letter to the First Lord of the Admiralty, Lord Tweedmouth, in February 1908, failed to alleviate British fears. It was left to the King to send a polite but sceptical reply.

It all boiled down to the fundamental difference in the strategic positions of the two rivals. Britain alone could not invade Germany, which had a short coastline and a supremely formidable army. Even a Copenhagen would not render Germany vulnerable to her Continental neighbours. Germany could even lose one or more land battles and still win a war. But if the Germans acquired a large enough navy to inflict one defeat on the British in a fleet action in the North Sea, Britain would lie open to invasion and risk losing her Empire. The British fleet was a matter of life and death; the German fleet was a virility symbol. The Kaiser simply refused to see the point and viewed every British suggestion to limit naval expansion as an insult designed to preserve the status quo. The rise of the High Seas Fleet thus radically altered British policy, to the detriment of the Second Reich. Eventually the resolutely maintained British naval supremacy strangled it.

But the British very much wanted a reduction in tempo. Lloyd George, then Chancellor of the Exchequer, was behind the three-to-two ratio proposal, having redefined the two-power standard as Germany plus any one other naval power. The British determination to stay ahead overtook concern about cost. Sir Edward Grey remarked: 'If the German navy ever becomes superior to ours, the German army can conquer this country.' Given the will of the British and their clear perception of the danger, the Germans might prolong the contest as long as they liked, but they could never win it. The Kaiser's vanity was very costly for his country. When he met Edward VII in Germany during August 1908, he simply refused to talk about Germany's naval build-up.

Two months later Austria made a dangerous intervention in the unstable Balkans by annexing Bosnia and Herzegovina, outlying parts of the Turkish Empire, in defiance of the Austrian agreement with Russia of 1897 to accept the status quo in the area. Tension had risen following the July revolution in Turkey; the annexation aroused alarm and anger in Serbia, Russia's independent southern Slav ally. Austria airily promised the Russians the right of passage through the Dardanelles for her ships, but was unable to deliver because of British and French opposition, the Triple Entente notwithstanding: they had not been consulted. In March 1909, however, the Germans demanded complete recognition from the Russians of Austria's annexations. The Russians did not feel strong enough to refuse and although on this occasion they complied it made them all the more determined to be in a position to show more resolution next time. Thus the Triple Entente was strengthened rather than weakened by Germany's intervention in a quarrel on the side of a weak ally who was now a liability.

The all-powerful German army made the situation worse. Moltke, the Chief of Staff, independently assured his Austrian counterpart, Hoetzendorff, of German military support against Serbia even if Austria provoked a war. The newly powerful navy was no less forthright in ignoring the subtler calculations of German politicians and diplomats, as the new Chancellor, Bethmann-Hollweg, soon found out when he tried to negotiate with the British for a political and naval settlement in autumn 1909. When the Chancellor complained to the Kaiser about interference by the naval attaché in London, the Kaiser decreed that an officer was immune to civilian interference.

The British did not respond to German overtures because they wanted too much for too little. The German demand for British neutrality in the event of a war between France and Germany was out of the question and was seen as a blank cheque for the Germans to swallow up Europe. The rise of the German navy made it impossible for the British to succumb to any temptation to return to splendid isolation. Mistrust of Germany was the key to the British attitude, anxious though they remained for a genuine rapprochement.

The Germans found their hands tied by their commitment to ailing Austria, which now had sufficient strength only to cause trouble. They saw clearly enough that Austrian involvement in the Balkans might easily embroil them with no possibility of benefit to Germany. Their strategy in such an event had long since been dictated by the army in the 1890s. In the event of conflict with Russia, Germany would first attack France to knock her out of the war and thus secure the rear for a protracted struggle in the east. French static border defences would require a German outflanking movement through Belgium. This was the Schlieffen Plan, drawn up by Moltke's predecessor, and it lay behind the demand to Britain to remain neutral. The British were not fully aware of these considerations, but their instinct led them to stick with France and thereby Russia. The Kaiser's extraordinary exclusive interview with the *Daily Telegraph* in October 1908 made this easier. Wilhelm told the newspaper's correspondent that he was personally favourably disposed towards Britain, but that the German people were hostile. The talks at the end of 1909 led nowhere, breaking down mainly on the naval issue, on which the Germans refused to be pinned down throughout the year as their capacity to build more capital ships grew rapidly.

In Britain public feeling on the Royal Navy's strength and German rivalry was highly volatile and came close to panic. The decision to build eight dreadnoughts was a response to that public opinion, but also had the effect of making the Germans pause in their own programme. Italy and Austria were building four each, while France concentrated on submarines and Russia was doing very little to reconstruct her fleet after the Japanese defeat. All this strengthened British resolve to stay ahead. The naval estimates for 1910 called for another five capital ships to give Britain twenty-five dreadnoughts by 1913, by which time the Germans might achieve twenty-one.

If the Kaiser had looked objectively at the situation he would have seen that the building of the German fleet produced the opposite effect to that desired: it brought Germany no allies and served to strengthen the Triple Alliance by bolstering British commitment to it. In the 1909 talks, the Germans suggested a four-to-three ratio in exchange for what amounted to a free hand in Europe (and the chance to dispose of Britain after the agreement expired). The British wanted a naval agreement as a pre-condition to a political one; the Germans wanted the converse.

The Liberals survived two general elections at the beginning and end of 1910, a year which saw the transfer to home waters of warships from Australia and New Zealand, and an important increase in the French naval programme.

Early in 1911, Grey, the Foreign Secretary, proposed to the Germans an exchange of information on current and planned naval construction. In March, the Germans responded with a complicated counter-proposal which would have had the effect of putting the British in the position of determining their naval programme without reference to Germany's. General talks between the two governments continued and appeared in June to be going well after agreement on simultaneity in concluding naval and political agreements. But the British refusal to concede neutrality in exchange for a German naval slow-down produced deadlock. Negotiations on an exchange of naval information fell away. The rest of the year was dominated by a new and highly dangerous confrontation between Germany and France over Morocco.

The Algeçiras agreement of 1906 had broken down in the following year because of internal unrest in Morocco and the weakness of the Sultan. Negotiations between France and Germany led to an agreement in February 1909 on equality of commercial and financial interest in Morocco, while Germany recognised France's special political position there. Early in 1911 the new Sultan's financial difficulties, and the taxes he imposed to deal with them, led to a revolt which occasioned the French to send an expedition to Fez in April to protect European lives and property, with a promise to withdraw when order was restored. The Germans took exception to the French unilateral initiative, while the French suspected that the Germans planned to exploit the unrest in Morocco to extend their influence there into the political sphere. Secret Franco-German talks of which the British knew nothing took place, during which the Germans made major demands for cession of French African territory. The cuckoo was still hungry, feeling as strongly as ever that it had been done down in the European scramble for Africa in the later half of the nineteenth century. The French found themselves in an impossible position: a little local difficulty suddenly became a challenge to their status as a world power. Then in July the Germans sent the gunboat *Panther* to the Moroccan port of Agadir to 'protect German interests'.

The British Government was thoroughly alarmed and even issued public warnings; the German press succumbed to Anglophobia. There were rumours of secret naval movements. A war-scare struck financial circles in Berlin; in London, insurance premiums against the risk of war at sea were doubled. The Royal Navy started moving coal experimentally from South Wales to Scotland by rail in September and alerted its North Sea squadrons. The Kaiser personally requested another naval increase. The Entente Cordiale, however, once again held firm. The Germans, who had originally demanded the whole of the French Congo, settled for a relatively small strip of territory in French West Africa and recognised (again) French political predominance in Morocco.

It was a small enough gain for the Germans at enormous risk, but the Kaiser and Tirpitz exploited it at home by agreeing to go for a two-to-three ratio in capital ships vis-à-vis Britain. The Admiral's blind devotion to his risk theory was as strong as ever, and he was encouraged by apparent British acceptance of the ratio, as shown by the relatively modest naval estimates of 1911.

Over the turn of the year there was a prolonged wrangle in Germany about the scale of the latest naval expansion plan. Tirpitz wanted to add three battleships and three battlecruisers to existing plans for the next six years. The Chancellor, the foreign ministry, the government's financial advisers and the Generals showed unprecedented opposition. The finance for a single dreadnought cost at least as much as raising an entire new army corps, which might well have been of more use to Germany at this stage, as the war clouds were gathering and the navy was surely powerful enough already. The Social Democrats gained a million votes in the election in January 1912 and achieved one-third representation for the first time: militarism and naval ambition were neither universal nor incapable of reconsideration. In his Speech from the Throne in February 1912, the Kaiser compromised by announcing the addition of three battleships to the programme of twelve capital ships for the period to the end of 1917. The British were not appeased; the French meanwhile adopted a programme, giving them twenty-eight capital ships by 1920.

At this time leading British and German businessmen were in touch with each other about the steady deterioration in Anglo-German relations. They hoped to create a climate of opinion among their government contacts which would lead to a renewal of political exchanges and a reduction in tension, which was bad for business. Winston Churchill, who had become First Lord of the Admiralty in October 1911, said Germany would have to reduce its naval programme before anything positive could happen, while the Kaiser said that he was willing to talk, but reserved the right to reinforce both the navy and the army in the meantime. The British were prepared to talk about colonial concessions and a mutual non-aggression agreement if only Germany would withdraw from the naval competition. In the end they were prepared to accept the latest German naval increase if it were spread over the next twelve years instead of six. Then the Germanophile Lord Haldane went to Berlin at the Kaiser's invitation and with the reluctant blessing of the British Cabinet, which gave him a brief to listen but not to negotiate, something the Germans completely failed to grasp or accept. Immense misunderstandings ensued and nothing came of the contacts. In March, as the British fleet was being reorganised for home defence, the Germans crossed the Rubicon by publishing the text of the supplementary Fleet Law foreshadowed by the Kaiser in February's Speech from the Throne. Churchill publicly proposed a 'naval holiday' in March.

This was no empty gesture. By this time the Royal Navy's lead over Germany and America combined in terms of capital ships amounted to 13 per cent, but the margin had fallen by 25 per cent in three years and the lead in completed dreadnoughts was just 4 per cent. Allowing for the German and American building programmes and also for the retirement of old ships, Britain would have to lay down seven dreadnoughts in 1912 to achieve a 10 per cent margin in 1915. The Admiralty at this stage made it clearer than ever that Germany was its over-riding concern by redefining the essential superiority standard as Germany-plus-60-per-cent in dreadnoughts, a ratio of five to three. Churchill announced estimates for the building of four capital ships, eight light cruisers and twenty destroyers, with the warning that this would be supplemented if necessary. The Americans were henceforward left out of the calculations, which indicated that if Germany did adhere to two keels a year for the next six years, Britain would achieve the desired 60 per cent lead by laying four and three a year alternately over the same period. If Germany added three more, as threatened, Britain would start six; if Germany reduced, Britain would too. The British adhered to a rather higher margin of superiority in lesser warships. Churchill said Britain would have to build five big ships in 1913 because Germany would build three: 'Supposing we were both to take a naval holiday for that year?' Germany, by not building three ships, would reduce the Royal Navy by five super-dreadnoughts (the second generation was already in hand), which, remarked Churchill in a typically forceful piece of rhetoric during his speech on the estimates, would be rather more than she could hope to do in a brilliant naval action.

Four days after Churchill announced his estimates, Tirpitz disclosed Germany's, adding three battleships to the programme laid down by the two Fleet Laws and their amendments, which had a target completion date of 1920. The Navy League, which by now had more than a million members (its magazine, *Die Flotte*, was selling nearly 600,000 copies an issue), thought this was not enough, but the Bill went through, despite concern about finance: the army was also to be reinforced. Churchill duly announced in May that a supplementary naval estimate would follow in two months. Over the next five years, Britain would substitute a start-rate of five in the first and four annually thereafter instead of three-four-three-four-three, an overall increase of four dreadnoughts. There would also be additional smaller ships and a recruiting drive. Austria had also started a secret, but hardly invisible, dreadnought programme; and in 1912 Russia announced her decision to build a new battle fleet at a cost of £50m (she had preferred since the 1905 war with Japan to rely on defensive torpedoboats). In the autumn of 1912, after the British had reorganised the navy to concentrate forty-nine battleships in home waters to the Germans' available twenty-nine, the French announced that they would in turn concentrate their fleet in the Mediterranean. The mutually highly advantageous Anglo-French arrangement, which finally destroyed the German risk theory, was effected in a non-binding exchange of letters between the London and Paris governments. It might just as well have been binding, for the double redeployment stripped the French Atlantic coasts of French ships and only the British could now protect them with the one navy that could face the Germans. To cover the

Empire, the British Dominions and colonies had since 1909 been increasingly helping with their own security by buying extra ships.

In October, war broke out in the Balkans, with Serbia, Montenegro, Bulgaria and Greece fighting Turkey. An armistice was declared in December and peace talks opened in London, but in January 1913 the Bulgarians attacked again and hostilities were resumed. By May the major powers were able to impose a territorial settlement in the region, and the crisis management committee of Ambassadors to London chaired by Grey, the British Foreign Secretary, brought the Germans, the French and the British together in a rare spirit of co-operation, which surprised and pleased all concerned. All parties were in fear of a spread of the fighting across Europe. Hope rose for a time that this new atmosphere would lead to a reduction in tension between Germany and Britain and an improvement in relations at last, even though the Germans had announced in March and April further increases in their army and navy, and new expenditure on aeroplanes.

In June, Bulgaria rounded on her anti-Turkish allies but was defeated in a matter of weeks. Behind the unrest in the Balkans lay the weakness of the Turks and the Austrians, and the Slavic sympathies of the Russians. The Austrians and their allies, the Germans, believed that a general European war over the Balkans was inevitable. Not only was weak Austria becoming more and more involved in difficulties with the southern Slavs, particularly the expansionist Serbians; the Germans also took the view that a final settlement of accounts between themselves and their eastern neighbours, the Russians, had to come sooner rather than later. France was tied to Russia, and now Britain was tied to France. The Germans therefore built up their army to be ready for the swift blow against France to clear the decks for a long struggle with Russia; there could be no question of relaxing the pressure on the British, whom they still hoped to separate from the Triple Entente, by naval concessions. It was not in the Kaiser's nature to substitute the carrot for the stick, the only move which might conceivably have detached Britain. It was too late now anyway. As late as July 1914 Grey retained the hope, aroused by the spirit of the Balkan peace conference, that the crisis-management technique developed there might also work in a larger context; but the co-operation of 1912 turned out after all to be brief, limited and superficial.

Just before Churchill was due to announce the naval estimates for the financial year 1913–14 in March 1913, Tirpitz told the Reichstag that he was prepared to accept, in terms of battleships only, a British margin of superiority of 60 per cent in dreadnoughts, or in terms of ships, thirty-two to twenty, and in squadrons eight to five. This was no more than a clever ruse to gain time because it took no account of heavy cruisers or of the acceleration involved in the 1912 supplementary Fleet Law; but the Germans were also anxious to divert funds to a major expansion of the standing army by 25 per cent in the next three years. The navy simply had to take second place, at least for a year – and there was already talk of a third generation of dreadnought with 15-inch guns and a displacement of 40,000 tons each. Churchill dismissed the Tirpitz concession as hollow and not a proper answer to his renewed offer of a year's naval holiday: the German proposal had been expressed strictly in terms of an agreed level of eight battleship *squadrons* of four apiece to Germany's five similar, and not in terms of cuts in construction plans. Despite nationwide Liberal agitation against the costs involved in shipbuilding (£45m in 1912 and £49m in 1913), Churchill announced five new dreadnoughts, eight light cruisers, sixteen destroyers and 7,000 men. The Admiralty was now aiming at a three-to-two lead in home waters over Germany, with the remaining 10 per cent of the 60 per cent margin for use in training and missions further afield. The growing contribution in ships from the Empire was not counted because it was intended for the defence of the Dominions and colonies. But when the Canadian Parliament threw out a Bill for the construction of three dreadnoughts, Churchill committed Britain to building three more if Canada did not change her mind in 1914, and he was confident they could be included in the 1914–15 estimates.

Meanwhile, the risk in relying on battleships was growing apace, which is why thought was being given to even bigger guns with longer range. The effectiveness of the torpedo had risen enormously since the Russo-Japanese war, thanks to a five-fold increase in range to 11,000 yards. Since 1913 the Germans had been busy increasing their strength in the air for reconnaissance purposes by building airships and aircraft, including seaplanes which could be launched from and recovered by capital ships. It was going to be much harder to hide ships in the confines of the North Sea.

When the Kaiser wanted to start the third extra dreadnought, provided for in the 1912 supplementary Law, as early as 1915 and to introduce another supplementary Law to step up the number of cruisers for use in distant waters, Tirpitz, for the first time since 1898, had misgivings about the expansion of his navy and opposed his monarch. In February 1914 he wrote to the German naval attaché in London that another increase in the German rate of construction would be 'a great political blunder' which Churchill would use to British advantage. Today this would be called a U-turn; in this instance, it was an extraordinary admission of defeat prompted by financial stringency. Germany could no longer stand the pace – but only after spending more than £200m on warships between 1900 and 1914, without getting out of the danger zone Tirpitz's risk theory had called into being.

As the war-clouds turned from grey to black, soundings were still going on about reducing armaments among the powers who had taken part in the abortive first and second Peace Conferences in The Hague in 1899 and 1907. In May 1914, the American President Wilson sent his personal emissary and plenipotentiary, Colonel House, on a tour of the major capitals to discuss arms limitations. The special adviser proved unequal to the task, not only because of his ignorance of the complexities of European politics but also because of the growing resignation that war was inevitable and the general appetite for it, which was growing by the hour. Anglo-German naval rivalry was a principal cause of the First World War. No other cause surpassed it in importance. The Tirpitz Memorandum of 1894 was to prove the undoing of the Kaiser.

The final crisis blew up in the Balkans to destroy the illusory Anglo-German detente which had begun there less than two years earlier. In June 1914 assassins killed the Austrian Archduke Franz Ferdinand and his wife at Sarajevo. The Germans hoped that the British would not after all go to war on the Continent over the Balkan situation, especially as there was a threat of civil war in Ireland over the Home Rule issue. Britain, if pressed, might well admit to being unable to contain her indifference about Serbia; but King George V and Sir Edward Grey gave clear warnings that Britain could not stand aside if war broke out between Germany and France. Russia had regained enough strength and confidence to stand up to Austria over Serbia, even when Germany backed Austria with an ultimatum. France was fully committed to Russia. Germany was committed to the Schlieffen Plan. Britain asked Germany to restrain Austria from attacking Serbia, without avail. The Germans felt pressed for time because, given their belief that war was inevitable and given also that the French had heavily reinforced their standing army, they wanted to deal with the Russians before they had fully recovered militarily (estimated for 1917). Elsewhere the widening of the Kiel Canal had just been completed, enormously increasing German naval mobility by enabling them to reinforce in the North Sea from the Baltic, but Tirpitz felt that the navy, after its enormous expansion which had quadrupled manpower in seventeen years, was not yet ready. The Royal Navy mobilised at the end of July as Grey committed Britain to support France; the Russians mobilised their armies; the Germans swallowed their last-minute misgivings and followed suit; Germany declared war on Russia at 5 p.m. on 1 August 1914 and on France on 3 August. The British Empire came in the next day at midnight. The Schlieffen Plan was not to be denied.

It remains to record the strength of the two navies which faced each other across the North Sea when war broke out. The figures vary somewhat between one authority and another, difficult though it is to accept that anyone could overlook or imagine ships as large as dreadnoughts. Some sources, however, include ships that were not quite ready but very soon would be; others include ships that were not quite dreadnoughts, or not quite British. The Royal Navy requisitioned two dreadnoughts just completed for Turkey. With these and the last two pre-dreadnoughts of the Lord Nelson class (same speed, fewer guns), Britain had twenty-four dreadnoughts afloat and another thirteen under construction. Germany had thirteen in commission and ten under construction; two were due for completion at the end of 1914 in each country. Of invincibles, Britain had nine plus HMAS *Australia* while Germany had five plus SMS *Blücher*, their under-powered and under-gunned first attempt at a battlecruiser, which proved to be more of a liability than an asset. Including the foregoing, Britain had forty-three battleships up to fifteen years old to Germany's thirty-one, and thirty-four heavy cruisers to Germany's eight. Britain had thirty-six light cruisers to Germany's thirty. There were 143 modern destroyers in the Royal Navy to 127 torpedoboats in the Imperial Navy (the two terms reflect a linguistic rather than a functional difference: they were all torpedoboat-destroyers), with thirty-six and twelve respectively on the stocks. Of submarines, Britain had thirty-seven plus twenty-nine on order to Germany's twenty-seven and fourteen. Such was the result of the Anglo-German naval arms race when the war it had done so much to bring about finally began.

PART II

Willing to Wound but Afraid to Strike – the War at Sea

There seems to be something wrong with our bloody ships today.

– Vice-Admiral Sir David Beatty at the Battle of Jutland, 31 May 1916

Our country needs to care for nought:
The Fleet is fast asleep in port!

– street-chant in Wilhelmshaven after the Battle of the Dogger Bank, 1915

CHAPTER THREE

The Skirmishes Before Jutland

When war broke out the expectation on both sides was that it would be short. The standing armies entrained or marched to their border assembly areas in the confident assumption that it would all be over by Christmas. The Germans remembered their swift defeat of France in 1870, took the view that war had not changed so radically since, concentrated their main forces in the west and looked forward to a repetition, which would leave them free to deal with the Russians. The vastness of Russia needed more time, but Germany, the world's leading military power, did not believe that the Russian army posed a serious problem. They could not have known how right they were: it took just three years to destroy the Russian Empire and impose the harsh Treaty of Brest-Litovsk on the Revolutionary Government which ousted the Tsar, even though the Germans were simultaneously locked in the mire of the Western Front in huge numbers. Militarily, the Schlieffen Plan was the right strategy, and it came close to success despite the most appalling blunders. Politically, however, it could be justified only by early success. To outflank the fixed defences of the French, the German right wing had to march through neutral Belgium and Luxemburg, which aroused foreign opinion against them. The French were also a different proposition compared with the ineffectual Second Empire. They wanted their lost northern provinces back, and this time they were not alone. Yet what did the Germans have to fear from the Belgian army or the British one, an all-regular force of just 150,000 men which had to be ferried across the Channel?

Apart from their early blunders in the field, deriving mainly from failures in co-ordination and communication, the Germans had failed to appreciate that weapons of defence were more effective at the time than those of offence. This applied at sea as well as on land. The machine-gun, the wire and the trench withstood the artillery barrage and the bayonet charge; the mine, the torpedo and the submarine restrained the surface fleets. The war ceased to be one of movement and became bogged down, a total war between peoples and not a decisive encounter between professional forces.

Let us now turn our back on the land, where the dying took place, and focus on the sea where the war was won. The generally expected short, sharp war offered the Royal Navy only one possibility of influencing the result: a Trafalgar against the German navy. The British hoped for this in the early days, but the Germans were not about to oblige them. They expected the British to adopt their traditional naval tactic of blockade and drew up their own tactics with this in mind. They planned to pick off individual ships or formations in the blockade line by concentrating superior force against them, until Britain's naval superiority in numbers had been reduced by attrition to such an extent that the High Seas Fleet could come out in strength and take on the British Grand Fleet on even terms.

The German navy was extremely well placed to do this. The North Sea coastline was short and well shielded by islands and tortuous channels through the sandbanks. Neutral neighbouring territory (the Netherlands and Scandinavia) prevented outflanking. The sheltered Kiel Canal, wide enough for the latest capital ships in 1914, provided a back door to the Baltic. The strongly fortified island of Heligoland provided the main North Sea bases of Wilhelmshaven, Cuxhaven and Bremerhaven with strong forward defence. All this made an excellent basis for defence against attack and for raiding operations. That extraordinary and prophetic thriller, *The Riddle of the Sands*, written by Erskine Childers, a clerk to the House of Commons in 1903, describes as nothing else could the nature of the German North Sea coast and its value to the German navy.

Admiralty war plans did indeed envisage a close blockade of Germany to prevent raids and to spot sorties by the High Seas Fleet so that it could be 'brought to action by the British Main Fleet'. But a close blockade meant capturing the German islands, which were heavily fortified, while U-boats and torpedoboats could wear down blockading ships. Thus the close blockade strategy was abandoned in 1912 in favour of an 'observational' blockade, a line of ships in a 300-mile curve from Norway to Holland, with the Grand Fleet lying in wait to the west of it. This would, however, have been both difficult to make effective and an invitation to the Germans to concentrate superior firepower at will against individual ships.

On the very eve of war, in July 1914, the Royal Navy substituted the 'distant' blockade, well clear of minefields, hit-and-run raids and the possibility of piecemeal diminution. The Channel Fleet would cover the Straits of Dover and the Grand Fleet the 180-mile gap between the Orkney Islands and Norway from bases in Scotland. Thus the two exits from the North Sea to the Atlantic would be closed to Germany, and Britain would make absolute domination of the North Sea the basis and priority of her naval strategy. This proved to be the key to victory. It was neither heroic nor glorious; it was, nevertheless, 'the pivot of the Allied cause', as Sir Basil Liddell Hart comments in his *History of the Great War*: 'Its effect on the war was akin not to a lightning flash, striking down an opponent suddenly, but to a steady radiation of heat, invigorating to those it was used in aid of, and drying up the resources of the enemy.' Liddell Hart identifies the decisive moment of the war at sea as having occurred on 29 July, six days before Britain even declared war. As luck would have it, the Grand Fleet had gathered in Portland for a review on 26 July. It was ordered not to disperse because war seemed imminent, and on 29 July it sailed as a body to its wartime anchorage in Scapa Flow in the Orkneys, unremarked: 'From that moment Germany's arteries were subjected to an invisible pressure which never relaxed . . .'

The Germans expected the British to apply a close blockade at the start of hostilities to counter any early move of their own, and also to cover the transport of the British Expeditionary Force to France. They hoped this would give the High Seas Fleet the chance to dash out and overwhelm isolated British squadrons. Even when the implications of the distant blockade became clear, the basic German strategy never changed: no pitched battle with the Grand Fleet until it had been sufficiently whittled away by attrition.

On 6 August, the Channel Fleet, led by pre-dreadnought battleships under Admiral Sir Cecil Burney, covered the tranquil crossing to France of the first four divisions of the BEF. The entire Royal Navy was on maximum alert. The Grand Fleet swept the North Sea, the Harwich Force of destroyers led by light cruisers and supported by submarines covered the southern North Sea and the Dover Patrol of destroyers and light craft covered the Channel Fleet. The two destroyer-formations had been specially assembled for the war and were to play a disproportionate and gallant role in it. Wise though this immense naval operation was, it found almost nothing to shoot at. The British were astonished that the Germans made not the slightest attempt to interfere with the troop landings, soon to be followed by more. A handful of German submarines and fast patrol-boats could have caused havoc by hit-and-run raids among the troop transports. Even the over-confidence of the German General Staff does not fully explain the neglect of an opportunity to cause maximum confusion at minimum cost. The effect on enemy morale would have more than justified the risk. It was an oversight as inexplicable as the German failure to destroy the BEF at Dunkirk in 1940. Perhaps the same vague hope that the British might still have made a separate and early peace played a role in it. Throughout the war, the gigantic ferry service to and fro across the Channel remained inviolate, shielded by a floating iron wall at which the Germans could direct only ineffectual, if sometimes spectacular, pinpricks.

The German General Staff, to which the navy was directly subordinated, undoubtedly expected a swift victory in France, so swift that whatever troops the British might be able to send across would be irrelevant. But the inactivity of the German navy in the early days of the war and in the weeks immediately preceding its outbreak deprived the Germans of a whole range of other opportunities for harassment. Britain was totally dependent on the sea for survival, but remarkably few commerce raiders were deployed by the Germans. The successes of those they did use were short-lived but showed how highly damaging they could have been in greater numbers. The converted luxury liner *Kronprinz Wilhelm* (not to be confused with the battleship of the same name) caused chaos in the North Atlantic after being painted grey and fitted with two popguns borrowed from the navy; this was as nothing to what the detached cruiser *Emden* did in the Indian Ocean, as will be seen. The Germans' own anticipation of a British blockade of some sort makes the missed opportunity all the more puzzling, even if the Germans faced serious difficulties in coaling in distant waters. But then nobody on either side thought it would be a long war. When that became clear, the Germans turned to the submarine and almost won the war with it.

The destroyer HMS *Lance* had the honour of firing the first British shot in the war, at the minelayer *Königin Luise*, caught red-handed off the mouth of the Thames on 4 August. The German vessel scuttled itself. It had been repainted to look like one of the British railway steamers on the Harwich–Hook of Holland run (it was an excursion steamer before the war). The *Lance* therefore went on to open fire on a similar-looking ship which was flying a white flag of truce and was spotted in the same area. This turned out to be a genuine British railway steamer taking the expelled German Ambassador to London and his staff home. The *Lance*'s flotilla leader, the light cruiser *Amphion*, had to interpose herself to save the ferry. On the way back to Harwich, the *Amphion* struck one of the mines laid by the *Königin Luise*, blew up and sank with the loss of 169 lives, including eighteen Germans rescued from the minelayer.

German mines were far superior to British: before the war the Royal Navy neglected the weapon, regarding it as superfluous and distasteful, while the Germans had realised its potential.

The submarine was an untried weapon of very limited capacity and performance in 1914. But both sides had already recognised its value for reconnaissance at least, and began to put the initially primitive boats to use from the very beginning; the British to spy in the Heligoland Bight area and the Germans as a defensive screen and a means of locating the apparently idle British fleet. In the latter task, the First U-Boat Flotilla of ten submarines suffered the first casualties in this form of warfare. One sank by accident, one broke down and a third was spotted off Scapa Flow in November 1914 by the light cruiser *Birmingham* of the Grand Fleet, which rammed and sank her. The ever-careful John Jellicoe, who was appointed commander of the fleet the day war broke out, was worried about the potential submarine threat from the very beginning and moved his vast force ceaselessly in the opening weeks. Scapa Flow, his principal wartime anchorage, was the perfect deep-water base but at the beginning of the war had no defences against submarines.

A skirmish in the southern North Sea between a German reconnaissance group of two light cruisers and two submarines, and a British light cruiser and sixteen destroyers on 17 August, was not pressed to a conclusion by either side. The British next decided to launch a destroyer attack on the Heligoland Bight. Two light cruisers and thirty-one destroyers of the First and Third Flotillas, backed by six more light cruisers, started the raid on 28 August. Meanwhile, to distract the German navy while Royal Marine reinforcements were being ferried to Ostend to protect the Belgian port from impending German attack, the Admiralty decided to enlarge the raid, by including a squadron of five battlecruisers under the command of David Beatty. The original light force was not told because radio silence had been imposed, and suffered a nasty shock when the capital ships appeared unannounced. A catastrophic misunderstanding was narrowly avoided, one of several breakdowns in communication which materially reduced the effectiveness of the raid and first brought to light one of the principal shortcomings of the Royal Navy throughout the war – inefficient signalling. Throughout the Battle of the Heligoland Bight, the first major naval exchange of the war, the British ships spent as much time chasing their own tail as twisting the enemy's, wrongly identifying their own ships as German. Fighting began at first light, and by lunchtime one large German destroyer, the *V187*, had been sunk. The Germans could not bring out the three battlecruisers they had on hand because they were locked in behind the sandbanks before the mouth of the River Jade, which they could clear only near high tide around noon. Beatty intervened when the light British ships mistook a German light cruiser for a heavy one. The *Mainz*, the *Cöln* and the *Ariadne*, all German light cruisers, were sunk. Beatty ordered a withdrawal in the correct anticipation that German capital ships would be on their way as soon as the tide permitted them to come out. The British lost no ship, though one light cruiser was crippled and several destroyers damaged. It was a clear victory and first blood to the British, but hardly brilliant. German communications had also proved less than adequate. Commanders underestimated the scale of the British raid until it was too late.

After the battle, the Germans used great belts of mines to protect the Bight and felt confirmed in their policy of preserving their High Seas Fleet intact. Things were going well on land, after all. The Russian advance into East Prussia had been lured into a trap and then devastatingly repelled at Tannenberg; in the west, the great thrust through Belgium was driving all before it and the way seemed to be opening to a drive on Paris. An undamaged battle fleet would be a useful extra card to play in imposing an advantageous peace by Christmas.

It is striking how the early moves in the First World War at sea were to foreshadow those of the Second, including the unopposed transport of British troops to France, the initial underestimation of the submarine weapon, the importance of the smaller surface ships and the early bottling-up of the German navy. The British also had an enormous stroke of luck at the very outset of hostilities which was to be repeated in a different way (and surpassed) in 1939. On 14 August 1914 the German light cruiser *Magdeburg* sank in the Baltic at the mouth of the Gulf of Finland. The Russians found the German navy's cipher and signal codebooks and a master grid-chart of the North Sea clasped in the arms of a drowned warrant officer and passed them to the British, just as in 1939 the Poles imitated a German coding machine and handed it over. Thus for most of the early years of the war, the British were able to follow German ship movements closely. By the time the Germans became suspicious and altered their ciphers and signals, the British had developed the technique of radio direction-finding, and thus retained superiority in this new kind of intelligence-gathering.

Just what the Germans might have achieved on the open sea had they had the foresight to make the appropriate dispositions in the last weeks of peace was demonstrated in November 1914, when the Royal Navy suffered its only outright defeat in battle during this war, an event unprecedented in modern times and a shock to the British psyche.

The German East Asia Squadron under Vice-Admiral Maximilian Graf von Spee had been scattered in various parts of the Pacific at the outbreak of war, but by mid-October the main components were assembled at Easter Island. They were the modern heavy cruisers *Scharnhorst* and *Gneisenau* and the light cruisers *Nürnberg*, *Leipzig* and *Dresden*. Also there was a collection of colliers from which the ships refuelled, before they sailed for the coast of Chile. On 31 October, a Saturday, Spee, forty miles off Valparaiso, received a signal that the British light cruiser *Glasgow* was at Coronel, 250 miles south. By late afternoon on Sunday, 1 November 1914, Spee's squadron, with the exception of *Nürnberg*, was off Coronel. Two British ships were sighted, the *Glasgow* and the elderly armoured cruiser *Monmouth*. Shortly afterwards these were joined by the armoured cruiser *Good Hope*, which had been coaling in Port Stanley in the British Falkland Islands on the other side of Cape Horn, and the armed merchantman *Otranto*. These were the only ships at the disposal of Rear-Admiral Sir Christopher Cradock, who flew his flag in the *Good Hope*. In the evening the two battle lines were drawn up, the British to the west, with the temporary advantage of the sun behind them, and the Germans to the east, with their marked superiority in gun range. Cradock tried to close the range by steering for Spee but the Germans stayed clear. As the evening wore on, the grey of the German ships (livery invented by the Germans and introduced in 1902) merged into the twilight spreading from the east, while the British vessels were now silhouetted in the afterglow of the sunset. The Germans opened fire at a range of 12,000 yards with devastating accuracy. The two British heavy cruisers were soon damaged, their heaviest armament of older 9.2-inch guns reduced to impotence. Cradock turned towards the German line in an attempt to close the range for his secondary armament, but the Germans skilfully kept their distance and maintained their deadly barrage. The *Monmouth*, listing and on fire, fell out of the line. Then the *Good Hope* came under the concentrated fire of the two German heavy cruisers and exploded. The hulk drifted away and was never seen again. The *Glasgow* tried to help the *Monmouth*, which was crewed by reservists, but then was forced to flee south, and the *Otranto* had been ordered away by Cradock before the *Good*

Hope made its last charge at the German line. The *Nürnberg*, fresh from an errand and dashing to join Spee, came across the limping *Monmouth* and finished her off. There were no survivors from the two sunk ships, whose crews totalled more than 1,000 men.

The news reached London on the evening of 4 November. Churchill, the First Lord, and Lord Fisher, recalled to the service as First Sea Lord, decided on an immediate and overwhelming counter-stroke to try to make amends for what the British regarded as a disaster for morale and their unbeatable navy. The battlecruisers *Invincible* (the first ever built) and *Inflexible* set out from Devonport for the Falklands within a week. They were joined on the way by the armoured cruisers *Carnarvon*, *Cornwall* and *Kent*, and the light cruisers *Bristol* and *Glasgow*, which had been patched up in Rio de Janeiro after her escape from the Battle of Coronel. *Macedonia*, another cruiser, the *Orama*, an armed merchantman, and a flotilla of colliers joined the new squadron, commanded by Vice-Admiral Sir Frederick Doveton Sturdee. The British avengers reached the Falklands on 7 December. The following morning Spee appeared off Port Stanley. Sturdee was told while shaving, but coolly allowed his crews to take breakfast while his two battlecruisers completed their overnight coaling.

Incredibly Spee failed to take the only chance of reducing the formidable odds against him by attacking the British ships at anchor. He was in a position to inflict immensely disproportionate damage, though it would almost inevitably have meant death for his own ships. He sheered off to the south-east instead and the British came out in line ahead to give chase. The Germans opened up a twelve-mile gap. Spee ordered his three light cruisers, the *Nürnberg, Leipzig* and *Dresden*, to run for it while the two heavy cruisers took on the might of the enemy with extraordinary gallantry. The British armoured and light cruisers set off after the German light cruisers while the battlecruisers set about pounding the flagship *Scharnhorst* and the *Gneisenau*. The *Scharnhorst* was sent to the bottom with the loss of all hands, including Spee; the *Gneisenau* was reduced to a helpless hulk and scuttled herself after exhausting her ammunition.

The *Kent* caught up with the *Nürnberg* and sank her the same evening, while the *Leipzig* was despatched two hours later by the *Cornwall* and the *Glasgow*. The *Dresden* escaped and gave the British the slip until 14 March 1915, when she was cornered at Juan Fernandez Island by the *Glasgow* and the *Kent*. She ran up a white flag, put her crew ashore and scuttled herself by blowing up her main magazine.

Spee's East Asia squadron had been ordered to harass Allied shipping in the Pacific, but its achievements were nothing compared to the havoc caused by a single ship of his command, the modern light cruiser *Emden*, which he despatched to the Indian Ocean on detachment on 13 August 1914.

The mischief-making mission was the idea of the *Emden*'s captain, Commander Karl von Müller, and he was given the collier *Markomannia* in support. He passed through the neutral Dutch East Indies, and on the way added a false fourth funnel to make the *Emden* look like a British light cruiser. The British were already aware that the *Emden* had cut loose and were anxious to find her. She monitored heavy wireless traffic (a practice already in common use: at Coronel and the Falklands the two sides jammed each other's radio) and even picked up a cheeky (or desperate?) British signal which said: '*Emden*, where are you?' After being warned off peaceably several times by Dutch warships, the *Emden* sighted the immensely superior British cruiser *Hampshire*, but eluded her off Sumatra. On 8 September, the German warship seized the neutral Greek steamer *Pontoporus* because it was carrying British coal from Calcutta and put a crew aboard this godsend of a prize.

Müller then started work in earnest in the Indian Ocean, a British lake. On 10 September he seized and sank the *Indus*, an unladen cargo and passenger steamer, and on the next day a similar vessel, the *Lovat*. The crews were imprisoned on the *Markomannia*.

On 12 September he captured the *Kabinga*, carrying jute from Bengal to New York, and put all his captives aboard her under a German prize crew. During the night he boarded a British collier, the *Killin*, and the liner *Diplomat*, loaded with tea – both were sunk. The next afternoon the *Emden* met an Italian ship, whose captain, despite Italy's previous membership of the German-led Triple Alliance and present neutrality, broadcast *Emden*'s position. But the next evening she seized and sank the unladen British collier *Trubboch*, adding her crew to all the others on the *Kabinga*. This ship was abandoned with all the prisoners seventy-five miles off Calcutta that same evening, and the unharmed captives surprised the Germans by lining the deck and giving three rousing cheers as the *Emden* turned away. It was intended as a compliment, not an expression of relief, and a legend was born.

The raider's depredations continued. The next victim was the cargo steamer *Clan Matheson*. The well-practised *Emden* boarding party evacuated the crew, opened the seacocks and remembered to shoot the racehorses among the mixed cargo aboard before the ship went down. The crew were handed over to a neutral Norwegian ship the next day. The British captain saluted Müller for his courtesy.

After scouring the north-eastern waters of the Bay of Bengal without result, the *Emden* headed west and shelled Madras, setting fire to oil tanks and escaping unharmed. By this time fourteen larger warships of the British, Australian, French, Japanese and Russian navies were hunting the bold German raider. Müller went on to raise his total of merchant ships captured and/or sunk to twenty-three. He took a pause for rest, repairs and recuperation at the Indian Ocean island of Diego Garcia, where news of the outbreak of war had not yet arrived. By the end of October, the hunt was closing in. The *Emden* passed the Nicobar Islands on her way to raid Penang in Malaya, where she spotted and sank the Russian light cruiser *Jemtshug* by torpedo and gunfire. Heading out to sea, Müller sank the French destroyer *Mousquet*. The Penang raid behind her, the *Emden* slipped south-eastwards among the myriad islands of the Dutch East Indies, taking a two-day break for cleaning, running repairs and coaling. Müller then headed west to the British Cocos Islands to destroy the vital junction of undersea cables linking Australia, India and Africa at Direction Island. He anchored offshore on 9 November and put ashore a landing party, which demolished the relay station – but not before its operators were able to send the warning signal to the Allied searcher which was finally to put the *Emden* at bay. The Australian light cruiser *Sydney*, larger, faster and more heavily armed than the German ship, and acting as advance escort of an Australian and New Zealand convoy, answered the call within hours. The *Emden*, under way again, opened fire, and with the same high quality gunnery shown by the Germans at Coronel and the Falklands, which the British had found the time to admire, destroyed the *Sydney*'s fire-control system. The Australians fell back on their superior range and maintained fire, inflicting severe damage all over the German cruiser until, with all her main armament incapacitated, Captain Müller was forced to run her aground on a reef south of North Keeling Island in the Cocos archipelago. Eventually Müller ran up a white flag and for him and the survivors of his crew the war was over. The *Emden* lost 141 men and sixty-five were wounded; 117 survived unscathed the magnificent odyssey of chivalrous destruction which in three months had taken the *Emden* 30,000 miles to cause £15m-worth of damage, directing the efforts of up to eighty Allied ships at a time. Commander Müller became for the Germans an all but mystical hero whose splendid gallantry brought tears to the Kaiser's eyes and bore out his faith in his navy; for the British he was the nearest equivalent in the First World War to what Rommel came to represent in the Second. Müller, a model of a Prussian naval officer who combined stiffness and remoteness with dash, courage and courtesy, who preferred to capture and release rather than kill, who fought to the limit without losing respect for human life, was no less gallant than Spee's other captains who went down with their ships. One can only guess what might have happened had Spee dispersed his slender resources altogether and deployed the rest of his ships in a similar manner, or if the German navy had adopted a policy of hit-and-run raids on the open sea in sufficient numbers, positioning the necessary ships in anticipation of the outbreak of war. The German navy had a policy of replacing lost ships with similar vessels of the same name. The replacement *Emden* eventually became the flagship in internment of Admiral von Reuter from which he issued the order for the scuttling of the German High Seas Fleet. The two *Emdens* were exponents of the same tradition. Those of a romantic turn of mind will add that if there were a phantom ocean where dead ships sail, *Emden I* would salute *Emden II.* In fact both ships refused to go to the bottom when the end came.

In the North Sea, meanwhile, the German battlecruisers were occasionally being let off the leash. On 2 November 1914 they raided and shelled the coast of Norfolk at Great Yarmouth and on 16 December the north Yorkshire and south Durham littoral briefly came under their heavy guns. These raids had a nuisance value and were intended to probe British defences, and perhaps to catch isolated enemy ships or formations. Little damage was done, but the Germans escaped such pursuers as could be mustered. The British were naturally anxious to tackle these dangerous raiders and kept a close watch on German wireless traffic in the hope of correctly anticipating the next raid. On 23 January 1915 their diligence was rewarded when the radio monitors picked up a signal to Rear-Admiral Franz von Hipper, the German battlecruiser commander, ordering him to take his force from Wilhelmshaven and make a sweep down to the Dogger Bank area that evening. Once again there was no specific target; the operation was something of a speculative probe which might disrupt the fishing fleets of the British and yield an opportunity to attack naval patrols.

Hipper left the Jade estuary that evening with the First and Second Reconnaissance Groups. The First consisted of four battlecruisers, the *Seydlitz* (Hipper's flagship), *Moltke, Derfflinger* and *Von der Tann.* The last-named, however, was in dock for repairs, and her place was taken by the older, slower and less well-armed near-dreadnought *Blücher,* a large armoured cruiser. The Second Group consisted of four light cruisers. Each group was escorted by a flotilla of about twenty destroyers. Against them the British turned out Beatty's battlecruisers from Rosyth with the First Light Cruiser Squadron of four ships. Three more light cruisers and thirty-five destroyers were sent from Harwich as escorts. Beatty had reorganised his ships into two squadrons, the First, consisting of the *Lion* (his flagship), *Tiger* and *Princess Royal,* and the Second, the slower *New Zealand* and *Indomitable* under Rear-Admiral Moore, Beatty's second-in-command. But for the British this already superior force was only the first instalment. A further squadron of seven old battleships and another of four cruisers were despatched from Rosyth; destroyers and submarines were also sent on patrol; and Jellicoe led out the greater part of the Grand Fleet (three squadrons of battleships, three of cruisers, one of light cruisers and twenty-eight destroyers) from Scapa Flow. All these extra ships waited at a distance, ready to dash into a battle if one developed. The British had set a gigantic trap. On the unseasonally fine morning of 24 January, the opposing forces clashed when a German light cruiser fired on a similar British ship north of the Dogger Bank. Hipper soon realised he was outnumbered and in great danger and tried to evade the British. A chase developed. The German Admiralty responded to Hipper's report of the forces he had encountered by ordering all the High Seas Fleet battleships in Wilhelmshaven at the time to prepare for sea, a belated move.

Beatty's slightly faster ships began to gain on the Germans, whose speed disadvantage was exacerbated by the slowness of the *Blücher* at the rear of the battlecruiser line. The *Lion* at the head of the British line opened fire at the unprecedented range of 20,000 yards with her main armament of 13.5-inch guns. Soon the *Blücher* was taking savage punishment from three British battlecruisers. But the four German capital ships fired back and once more showed their marked superiority in gunnery. The *Lion* was soon considerably damaged by concentrated fire. A series of misreadings of signals, smoke, lack of initiative by subordinate commanders and poor gunnery began to erode British superiority. The German guns were smaller (11- or 12-inch) but better made and better controlled. The *Lion* was so badly damaged that she had to withdraw from the line and was towed home after the battle. The *Blücher* was crippled. The uninspired Admiral Moore misread a flag signal from the helpless Beatty which was in any case ambiguous (all other means of communication on the *Lion* had been knocked out), and concentrated the fire of all ships on the already doomed *Blücher* instead of trying to destroy the rest of Hipper's big ships, which took the chance of fleeing. The furious Beatty later boarded another battlecruiser to give chase, but by this time there was no chance of catching Hipper before he reached the protection of Heligoland and the High Seas Fleet.

It was a clear British victory and was hailed as such by a public thirsty for such news. The Royal Navy had sunk a heavy cruiser and seriously damaged the *Seydlitz* for the loss of fifteen lives (and the services of the *Lion* for four months). Nearly 1,000 Germans had been killed. But Moore's error, which let Hipper get away, dreadful signal readings, rigid adherence to orders rendered irrelevant by changed circumstances and some awful gunnery left the British with little to be proud of. Apart from the *Blücher*, the dead horse which was mercilessly and uselessly flogged at point-blank range, only three hits were made on the German battlecruisers, against twenty-two by the Germans on the British. More gunnery practice and accelerated installation of the latest 'director' co-ordinated firing system were ordered by the Admiralty, battle orders and signals were reworded to reduce confusion and more radios installed. On the basis of the damage to the *Seydlitz* the Germans improved deck armour against plunging shot; the British were to learn that lesson only at Jutland the following year. The Germans also improved the protection of their magazines against flash from enemy shells, another lesson the British had still to learn.

In terms of ships engaged, the Battle of the Dogger Bank was the biggest action of the war so far, but the outnumbered Germans had wisely run away to live to fight another day. German opinion was unimpressed by Hipper's prudence. The High Seas Fleet commander, Ingenohl, was replaced by Pohl, whose first contribution to the German naval strategy was to suggest an unlimited submarine campaign against enemy merchant shipping.

On 18 February 1915, the Germans declared British waters to be a war-zone in which all ships would be sunk on sight without warning. It was a ruthless decision akin to the violation of Belgian neutrality in the interests of strategy on the Western Front, and it blackened the German name. But once again it was the right decision in strategic terms, if not in political ones because of its effect on neutral America. It was also potentially lethal to Britain. The Germans correctly calculated that their U-boats could easily sink merchant ships at such a rate that the British would be unable to produce replacements fast enough. Pursued to its logical conclusion, the policy would starve the British out of the war. Only German vacillation prevented this happening, and even then it was, in Wellington's phrase, 'a damn close-run thing'. Unlike the Schlieffen Plan, the submarine campaign was foolproof, provided only that it was pursued *à outrance*. The first British response to this new threat was to tear up the 1909 Declaration of London which had imposed restrictions on blockade tactics. Now they assumed the right to intercept and arrest any ship on suspicion of carrying goods for the enemy and to take it to an Allied port to be searched. Consideration for neutral, particularly American, sensibilities had inhibited the British hitherto. Residual American hostility was removed when a U-boat torpedoed the liner *Lusitania*, on 7 May

1915, killing 1,100 people, including many Americans. The United States showed rather more understanding for the British after this catastrophe, which was to play a major part in bringing them into the war eventually. It took the Germans a mere six months from the start of their unrestricted U-boat campaign to reach the point where they were sinking merchantmen at a faster rate than the British could replace them. At the same time, Germany proved able to build three new submarines for every one she lost.

British counter-measures at first proved ineffectual to the point of impotence. The hundreds of patrol-vessels were hampered by lack of effective detection apparatus and weaponry. The Admiralty resolutely resisted the convoy system for far too long. Some merchant ships were armed, and for a while Q-ships – well-armed merchantmen with guns hidden under false upperworks acting as lures for U-boats – were able to sink a few boats; but they rather lost their point when the Germans ceased to identify targets before attacking them. But U-boats quite often stopped to pick up survivors if it was safe to do so. The campaign was undoubtedly against international law, but the Germans argued that British interference with shipping in their blockade was no less so and had the same aim: to sap the will to fight by starving the enemy population of food as well as war material. Yet the growing volume of American protests against the U-boat campaign led the Germans into the strategically fatal error of calling off the unrestricted campaign in British waters in September 1915 (it went on to April 1916 in the Mediterranean, where the risk of hitting an American ship was minimal), just as the U-boats had shown they could win the war. But the first counter-weapons were beginning to prove themselves, including primitive hydrophones for listening underwater and the first depth-charges for attacking submarines, as well as an experimental type of submerged bomb towed by fast vessels. But the British had signally failed to work out a defensive strategy or tactics against the submarine. The two methods used initially were inept: the offensive 'search and destroy' approach found few U-boats and sank fewer; and the defensive sweeps of shipping routes by patrols meant only that the Germans stayed submerged until the danger was past, and they could resume their attacks. From the start of the war to the Battle of Jutland at the end of May 1916, the Germans lost the derisory total of thirty-two U-boats, twelve of them by accident.

Fortunately for the British, once the unrestricted campaign was dropped, the German navy not unreasonably took the view that the risks attached to U-boat attack only after clear identification of targets was too high and submarine activity in the North Sea, except for highly effective underwater minelaying in the Channel by U-boats based on Flanders, virtually ceased. But the successes of the unfettered U-boats in the Mediterranean were a constant reminder of what could be done.

The British dominated the North Sea from the moment war began, and could thus fulfil their obligations to France without interruption. Their growing military contribution on the Western Front helped the French to hold the Germans, whose power on land was divided by the war on two fronts in which the Reich was engaged. With the Baltic closed to the Royal Navy, however, it was distinctly more difficult for Britain to help her other ally, Russia, especially as the naval situation in the Mediterranean was rather confused. The Germans had manoeuvred a divided Turkish Government into the war on their side, aided by events at sea. The British had commandeered for the Royal Navy two dreadnoughts being completed for the Turks on the eve of war; and the German battlecruiser *Goeben* with the light cruiser *Breslau* eluded the British and French navies in a dash across the Mediterranean to the Dardanelles, which they reached on 10 August 1914. Soon afterwards their Admiral was appointed commander of the Turkish navy and led a raid on the Russian Black Sea coast, shelling Odessa. Turkey was now inextricably involved on the German side.

Churchill cast about for a means of helping the Russians and opted for an assault on the Turkish-controlled straits of the Dardanelles, the narrow gateway to the Black Sea. On 3 November 1914, British and French ships shelled the forts at the entrance to the straits. As Russia wanted Constantinople and the straits for herself, she stupidly refused to co-operate. As Liddell Hart puts it, 'Russia would not help even in helping to clear her own windpipe. She preferred to choke rather than disgorge a morsel of her ambition.' The British and French ships bombarded the forts again in February 1915 and once more in March. Three of the eighteen Allied battleships were sunk by mines and the navy retired, calling for an army landing. On 25 April 1915, five Allied divisions (the British 29th division and the Royal Navy division, one French division and two of the Australian and New Zealand Army Corps – mostly raw troops) were landed at Gallipoli. By July the Allies had twelve divisions engaged to the Turks' fifteen (they had started with six under a German General). The invasion was a chaotic disaster, a sound strategic move wrecked by incompetent execution, and the Allied troops, after severe losses, were withdrawn in the three weeks to 8 January 1916, having achieved nothing.

Elsewhere in the Mediterranean, the Austrian navy dominated the Adriatic and the Germans made effective use of their best submarines against the long British lines of communication with their forces in Egypt and Gallipoli. Italy entered the war on the Allied side in May 1915 (although not against Germany until a year later), which helped the French to deal with the well-handled Austrian navy. By the end of 1915, despite a number of setbacks, the Allies had contained the Austrian fleet without destroying it.

But the focus of this story is the North Sea and the events that took place there and affected the outcome of the confrontation at sea between Germany and Britain. It is time to consider the one and only fleet action between the Grand Fleet and the High Seas Fleet.

CHAPTER FOUR

Jutland and the U-Boat Campaign

DESPITE THE GENERAL inactivity that followed the Battle of the Dogger Bank in January 1915 and although the British knew they heavily outnumbered the enemy at sea, the longing for a showdown (but only under the right conditions, which amounted to an advance guarantee of victory) persisted. Jellicoe did not need Churchill to remind him that he was 'the only man on either side throughout the war who could, by his actions, lose the war in an afternoon', as the statesman was to write later. The British had to keep naval supremacy at all costs, even if it meant winning by default; the Germans were no less determined to keep their fleet in being, if only as a bargaining counter for later use.

In February 1916 the ailing Admiral Pohl was replaced as commander of the High Seas Fleet by Admiral Reinhard Scheer, a much superior and markedly bolder officer. He assumed command when the vacillating Kaiser was in a mood to give his cherished High Seas Fleet more freedom of action, especially as the submarine campaign had been severely toned down. Scheer knew as clearly as anyone else that he could not take on the bulk of the Grand Fleet, but he felt he could be more daring in pursuit of the abiding German policy of luring parts of it to destruction piecemeal until a full-scale battle became feasible. He began to take the big ships out for brief sorties, always returning home before the Grand Fleet could get anywhere near. The British in turn tried to tempt Scheer into a proper fight by bombing Schleswig-Holstein from seaplanes. Soon after taking command, Scheer did come out in strength (twelve capital ships with escorts) but returned home early because of bad weather. Yet he had responded to the British bait, even if nothing had come of it. The Grand Fleet was not going to be drawn into action so close to the German coast, the only area where the Germans held the trumps.

Scheer next led the High Seas Fleet to sea on 24 April. The British Admiralty knew of the sortie, thanks to decoded wireless messages, and ordered the Grand Fleet to move out of Scapa Flow and the Harwich Force to put to sea. The Germans were in fact on their way to bombard the East Anglian ports of Lowestoft and Yarmouth on 25 April, in extremely remote support of the nationalist Easter Rising in Ireland, which the Germans had naturally done their best to encourage. The Harwich Force caught up with the Germans and tried to draw them southwards. But only the German Second Reconnaissance Group of four light cruisers turned aside to engage them, and forced them to withdraw. The German battleships lay off the north Dutch coast as the battlecruisers carried out the raid. Their commander, Hipper, was on sick leave and his deputy, Bödicker, withdrew at full speed after the brush with the Harwich Force, missing the chance the Germans had been waiting for to pick off a detached and temporarily isolated British formation. Scheer in turn decided to head for home. The two opposing main fleets had not come within 300 miles of each other. On 4 May the British staged another seaplane raid on a Zeppelin base in Schleswig-Holstein to try to lure Scheer out against the waiting Grand Fleet, but he did not emerge far enough and turned back early.

Scheer now planned a raid on Sunderland, chosen because it was much nearer to the Grand Fleet's base at Scapa Flow and closer still to Beatty's battlecruisers at Rosyth in the Firth of Forth, which were Scheer's real target. German submarines had been stood down from attacks on shipping in April after causing an outcry by sinking the steamer *Sussex*, and Scheer planned to place them off Scapa Flow and Rosyth to torpedo some of the British ships when they came to the relief of Sunderland. The submarines and also Zeppelins were to report British ship movements so that Scheer could get his timing right and withdraw after dealing with Beatty but before Jellicoe arrived. Jellicoe, meanwhile, had plans of his own to lure the High Seas Fleet to destruction by manoeuvring off Denmark and Norway.

Unexpectedly protracted repairs on the *Seydlitz* delayed Scheer by two weeks, by which time his submarines were approaching the end of their endurance. He abandoned the Sunderland raid scheme for a sortie up the west Danish coast by the battlecruisers with the battleships fifty miles behind, hoping that he could lure Beatty to attack his battlecruisers, use his battle squadrons to crush Beatty and withdraw before Jellicoe came. The British, aware in advance as usual that the Germans were planning a major movement for the end of May, told Jellicoe to shelve his own coat-trailing operation, scheduled for 2 June, until German intentions became clear.

Scheer sent the battlecruisers out at 1 a.m. on 31 May under Hipper, and led the battleships to sea ninety minutes later. The order to move for once defeated the British decoders, but they guessed a major deployment was imminent and Jellicoe was ordered to sea, as was Beatty, two hours before Hipper set off. The Germans steamed northwards in line, the British eastwards in two lines, the Grand Fleet battleships well to the north of the battlecruisers. For the first time in the war the opposing naval powers in the North Sea had committed the great majority of their capital ships to open water. Neither side was aware of the scale of the other's deployment, Scheer because his submarines underestimated the size of the British presence and his Zeppelins were grounded by poor weather, and Jellicoe because of a misleading signal from the Admiralty which falsely reported that Scheer's flagship, *Friedrich der Grosse*, was still in harbour.

The Grand Fleet (fleet flagship *Iron Duke*) deployed three battle squadrons, the Second, Fourth and First, each of eight battleships in two divisions of four apiece, and one battlecruiser squadron, the Third, with three ships. There were two squadrons of cruisers, the First and Second, each with four ships; one of light cruisers, the Fourth, of eleven ships; and three flotillas of destroyers, the Twelfth, of sixteen ships, the Eleventh, of one light cruiser and fifteen, and the Fourth, of nineteen, with two further destroyers attached to the battle fleet. Beatty's battlecruiser fleet consisted of two squadrons of battlecruisers, the First and Second of three and two ships respectively, plus *Lion* as fleet flagship, and one squadron of battleships, the Fifth Battle Squadron of four ships. There were three squadrons of light cruisers, the First, Second and Third of four ships apiece; two flotillas of destroyers, the First, with a light cruiser plus nine, and the Thirteenth, with a light cruiser plus ten; there were nine additional destroyers drawn from the Ninth and Tenth Flotillas; and finally, one seaplane carrier.

The High Seas Fleet (fleet flagship *Friedrich der Grosse*) mustered three squadrons of battleships. The Third had seven, the First had nine and the Second had six ships. There was also the Fourth Reconnaissance Group of five light cruisers (led by Commodore von Reuter); and another light cruiser leading three and a half flotillas of torpedoboat-destroyers, the First's number one Half-Flotilla, the Third, Fifth and Seventh. The battlecruisers formed the First Reconnaissance group of five ships, and were escorted by the Second, made up of four light cruisers. Finally, a further light cruiser led three more torpedoboat flotillas, the Second, Sixth and Ninth.

From this it can be seen how heavily the British outnumbered the Germans: Jellicoe's dreadnoughts alone outnumbered the German total of capital ships. The British had thirty-seven capital ships, all dreadnoughts, to the German twenty-seven (twenty-one dreadnoughts). The Grand Fleet without Beatty's battlecruiser fleet outweighed the whole of the High Seas Fleet including its battlecruisers. The British also had more and larger guns; each category of their ships outgunned its German equivalent, while the Germans' armour-plating was stronger, ship for ship. The German guns were more accurate, but the British fired heavier shells further. British dreadnoughts were faster than German, and had a much more elaborate gunnery control and direction system guided by a simple ancestor of the computer. The German disadvantage here was probably offset, if not actually outweighed, by the great superiority of their Zeiss rangefinders: getting the range right is after all the key to effective gunnery.

Completely unaware of each other's proximity, the British battlecruisers, heading east, and the Germans battlecruisers, leading the High Seas Fleet north, were about twenty miles apart at the moment when Beatty was due to turn north to join Jellicoe, at 2.15 p.m. on 31 May 1916. Both forces were screened ahead and alongside by cruisers and destroyers. Then chance, in the innocent shape of a neutral Danish steamer, took a hand. The *N.J. Fjord*, like a rabbit lolloping all unawares into the space between two packs of ravening wolves, reached a point as near as makes no difference to equidistant between the two battlecruiser formations, and was spotted by the two escort screens simultaneously. The British sent two light cruisers eastward to investigate; the Germans sent two westwards. At 2.20 p.m. the opposing pairs spotted each other, reported 'enemy in sight' to their respective flagships and opened fire. An hour later the five German and six British battlecruisers sighted each other, opening fire simultaneously at just over 15,000 yards (the British having thrown away their 4,000-yard range advantage by overestimating the distance between the opposing forces). Hipper, the German commander, was poised to execute the classic tactic of 'crossing the enemy's T' by taking his line across the bow of Beatty's leading ship and being able to concentrate his fire on it, the ambition of every line-of-battle commander. The German gunnery proved far more accurate far sooner than the

British, whose signalling once again went wrong in the heat of battle, causing dangerous confusion.

The British took at least four times as many hits as the Germans, who scored the first success. HMS *Indefatigable* was hit five times by brilliant salvoes from the *Von der Tann* and blew up, probably as the result of flash igniting her magazines. *Lion* would have gone the same way but for the presence of mind of a dying Royal Marine major who ordered the closure of the doors to the magazine when a German shell penetrated a gun turret. Beatty, having turned aside to parallel the German course and frustrate the crossing of the T, veered away and shooting ceased. Then his Fifth Battle Squadron approached Hipper's rear and opened fire at the unheard-of range of nearly eleven miles with 15-inch guns. Their shooting, at least, was as good as the Germans', whose guns could not reply at that range. Beatty turned back, heartened by the unduly delayed, but no less welcome, arrival of his battleships, which gave him a margin of nearly two to one. Hipper ignored the battleships and turned towards the British battlecruisers, the nearest enemy target. The action intensified. The battlecruiser *Queen Mary* exploded under murderous fire from the *Derfflinger* and the *Seydlitz*, again from flash igniting the magazines. The *Princess Royal* disappeared in a cloud of smoke after being hit, and was mistakenly thought to have met the same fate. Beatty remarked: 'There's something wrong with our bloody ships today,' and then ordered a change of course towards the enemy. Hipper had in fact turned away to sidestep a destroyer torpedo attack, which took the pressure off the depleted British line. Thus ended the first phase of the Battle of Jutland, with honours to the Germans and their bold and resolute battlecruiser commander.

Just over two hours after battle commenced, the British spotted the head of the line of Scheer's battleships to the south-east. Beatty's task was now to lure the whole German fleet north-westwards into the jaws of Jellicoe, who was pounding south-eastwards at top speed. It was a reversal of roles for the two well-matched battle-cruiser Admirals. Beatty teased and turned the Germans brilliantly until the two main fleets became aware of each other's presence and imminent clash.

Jellicoe's innate caution, combined with the distressingly usual signalling breakdowns and other errors, prevented the British from arranging their battle fleet in the way best suited to overwhelm the Germans and also stop them from running for home. Furious fighting between the smaller ships of the two fleets developed around the capital ships and several were badly damaged, more British than German. The British armoured cruiser *Defence* blew up. Shortly afterwards the *Derfflinger* and the *Lützow* opened fire on the *Invincible*, which was torn in two by a series of explosions. Each half tilted and sank in such a way as to leave bow and stern above the surface. Both fleet commanders were not surprisingly having difficulty in keeping their lines in some sort of order amid the confusion of battle, the distraction of subsidiary actions and high-speed manoeuvres. Scheer's line, eight miles long and sagging in the middle, was brought about in a brilliant manoeuvre under fire. Instead of going about in a great curve, or turning one formation at a time, each ship put about in turn, starting with the rearmost which thus became the foremost. The *Lützow*, which had, as Hipper's flagship, been leading the line, now became tail-end Charlie and the severe damage she had sustained led Hipper to transfer with his flag to the *Moltke*. Scheer was now heading west with his neatly turned line; he then executed another battle turn as immaculate as the first and headed back towards the Grand Fleet, which was in good order but only vaguely aware of the Germans' whereabouts.

There was about an hour of full daylight left. The German destroyers launched torpedo attacks which did little or no damage but disrupted the British line by forcing the big ships to take evasive action. Scheer ordered the battered battlecruisers into a last charge to help the destroyers at the head of his line. Then he sent the destroyers in again to cover another reversal of course as the British fire reached a new crescendo. The Germans had disengaged by the time night fell, and Scheer decided to run for home. But the Grand Fleet lay across his path as the second stage of the battle ended, shortly after 9 p.m.

As darkness fell, Beatty's battlecruisers, which had become detached from the Grand Fleet, opened fire on the head of the German line, the battlecruisers and the six pre-dreadnought battleships of the Second Squadron, separating them from the rest and forcing them to turn westwards, away from home. The British Second Battle Squadron, at the head of the main fleet, then caught sight of the silhouettes of the German dreadnoughts. Vice-Admiral Jerram did not attack, because he thought they were Beatty's ships, which he had been trying to find ever since Beatty had sent an impetuous message to the British battleships to follow him and help cut off the Germans. Scheer's battleships did not see Jerram's and the two lines turned away from each other without a shot. Neither fleet commander knew of the near miss. The British were particularly careless throughout the battle in failing to inform Jellicoe of sightings and their own or the enemy's movements, undoubtedly the major factor underlying the Grand Fleet's failure to nail the High Seas Fleet and drive home its advantage. This neglect of communication was in defiance of Jellicoe's detailed battle orders, which again and again stressed the importance of passing on information. The two fleets were now heading southwards, the British east of the Germans, about eight miles apart. Jellicoe sent his lighter ships to the rear in case Scheer decided to turn eastwards for home, ninety miles away, behind him. This is exactly what Scheer ordered. Hipper was at last able to get aboard the *Moltke* from a destroyer to which he had transferred from the abandoned *Lützow*, and he set out to take the four battlecruisers left to him from the rear, where he found them, back to their rightful place at the head of the German line. Only the damaged *Seydlitz*, down at the bow and listing after more than twenty shell and torpedo hits, gamely tried to follow.

The Germans were better equipped for night actions, which the British had never seriously practised. The High Seas Fleet had star-shells to light up the enemy, and searchlights which could blind his gunners. The cruisers and destroyers of the two sides skirmished inconclusively throughout the brief hours of summer darkness. The *Frauenlob*, a light cruiser, exploded and sank under British gunfire. The pre-dreadnought *Pommern* was hit by a British torpedo and also blew up, broke in two and went down with all hands, as the darkness began to fade.

Despite the many skirmishes during the night, Jellicoe was left to believe, until it was far too late, that the Germans were still to the west of him. In fact they had fallen back and turned across his wake for home. Only the four destroyers which led the attack that sank the *Pommern* sent back a clear message reporting the enemy's position and course. The Germans jammed it, but even had it got through clearly, it would have been too late. The Germans had passed behind the British battle line and forced their way through the tail of smaller ships, blowing up the cruiser *Black Prince* on the way.

The last shots of the Battle of Jutland were fired at 3.20 a.m. on 1 June by Beatty's separated battlecruisers – at a scouting Zeppelin. Jellicoe's hopes of renewing the battle and delivering a decisive blow finally died when a message from the Admiralty told him at 4.15 a.m. that Scheer's ships were already weaving their way through the German minefields off Horns Reef. The High Seas Fleet had escaped and despite casualties remained very much in being. Jellicoe realised he had missed the greatest opportunity of the war so far; nevertheless he retained total domination of the North Sea, which was his priority. All that was left to do was to look for German stragglers.

The argument about who won the Battle of Jutland, by far the biggest naval clash of the war, yet much less of an encounter than it might have been, has raged ever since. At least the statistics are known and clear. The Grand Fleet lost three battlecruisers, three cruisers, one light cruiser and seven destroyers; 6,097 men killed, 510 wounded and 177 captured. The Germans lost one pre-dreadnought battleship (*Pommern*), one battlecruiser (*Lützow*), four light cruisers and four destroyers; 2,551 dead and 506 wounded. These ships, all sunk, reduced the Grand Fleet by 111,000 tons or 8.84 per cent of its strength and the High Seas Fleet by 62,000 tons or 6.79 per cent. Tactically and morally, therefore, it has to be seen as a German victory, with the prize for boldness and dash going to Rear-Admiral Franz von Hipper. There was nothing to choose between the two sides when it came to courage, with a British boy-sailor winning a posthumous VC for keeping his gun firing and a German able-seaman winning the Iron Cross First Class for staying at the helm of his ship for twenty-four hours without relief and saving her. But surely the most remarkable feat was Germany's, to produce from nothing in twenty years a formidable navy which, ship for ship, proved to be a match for the British fleet with its centuries of glorious tradition and supremacy at sea. The Kaiser was as well served as the King.

Jellicoe had stuck to his agreed brief, which was to keep British dominance of the North Sea and the capacity to blockade Germany intact. Knocking out the High Seas Fleet was strictly secondary, but he failed to do it, despite some very astute manoeuvres which should have cut off the Germans from their bases, and would have done had Jellicoe been kept in the picture by his subordinates. He did not give the British the new Trafalgar they were thirsting for. The fact that his unglamorous strategy of distant blockade and supremacy at all costs, including forgoing a Trafalgar, won the war for the Allies was not quite enough. Prudence was part of his nature, and it was no doubt redoubled by the nature of his task. He might have been bolder at Jutland but for his unfounded belief that mines and submarines would be used against him, which was based on false intelligence.

Beatty was as dashing and as brave as Hipper, his opposite number, if not as efficient or as effective. He was also impetuous. Yet he supported his chief's caution consistently throughout the war, and out of conviction rather than loyalty. He might have made more of Jutland had he been in overall command – more of a victory, more of a defeat. As Grand Fleet commander, he was to follow Jellicoe's inglorious strategy of safety first to the end.

Scheer also had a primary and a secondary objective which mirrored and reversed those of the British. The first was to inflict severe damage on a detached section of the British fleet, which turned out to be the battlecruisers, the best he could hope for. In this aim he was unquestionably successful: he sank two out of Beatty's six and later one of the three directly attached to the Grand Fleet. Scheer's second objective was to keep his fleet in being by avoiding a confrontation with the bulk of the Grand Fleet. In these terms he failed twice; the second and third chapters of the battle involved both main fleets. On the other hand, he extricated his battleships on both occasions, even if it was more by luck than judgement, and brought them home safely, battered but unbowed. He too had fulfilled his orders; he too was governed by prudence and the determination to avoid unnecessary risk. As High Seas Fleet commander, Hipper was to follow Scheer's inglorious strategy of safety first to the end, or at any rate until it was too late for a change to make any difference.

Both sides claimed victory. The Germans' obvious case rested on numbers of ships sunk and took no account of the much higher degree of damage inflicted on the ships that got away, which made it impossible for the High Seas Fleet to put to sea again in a body for a good three months, whereas the Grand Fleet claimed to be ready to move again at four hours' notice the day after the battle ended. The British had twenty-four dreadnoughts undamaged to the Germans' ten.

The strategic victory belonged to the British, who retained undisputed mastery of the North Sea, and deterred the German

High Seas Fleet from ever again risking a fleet action (until desperation supervened). The indecisive battle of just twelve hours was decisive after all, precisely because it changed nothing. The British had far more to lose and less to gain than the Germans and did not lose it; the Germans had far more to gain and less to lose than the British and failed to gain it. The total destruction of the High Seas Fleet would not have affected the German war effort on land; a major reduction of the Grand Fleet would have rendered the British more vulnerable in their crucial command of the sea.

The Germans called it the Battle of the Skagerrak because that was where their battlecruisers won on 31 May; the British called it Jutland and celebrated one day later.

The fleet action in which Scheer became involved at Jutland was unintended and would probably not have happened had he been able to make full use of submarines and air reconnaissance. As far as he was concerned, his original plan, the raid on Sunderland to draw out *part* of the Grand Fleet for the High Seas Fleet to destroy by the application of locally superior force, followed by a rapid withdrawal before the rest of the British could join in, still stood; it was the only way Germany could hope to reduce the superior enemy's numbers to a level at which a real slogging match between the battle fleets could be contemplated.

The German Admiral therefore resolved to try again, and on 19 August 1916 brought a slimmed down fleet out once more. He left the five dangerously slow and less well-armed pre-dreadnoughts of the Second Squadron in port, but still had nineteen battleships, all dreadnoughts and all repaired at last after Jutland, and two battlecruisers (three were still not operational as a result of Jutland), plus light cruisers and destroyers. This time ten Zeppelins were up and four lines of submarines were in position off the English east and Dutch north-west coasts, to reconnoitre and report as well as to torpedo British formations. Scheer was confident that with so many watchmen on duty he would not this time encounter a larger

British force, but would be able to locate and demolish an inferior one and get away. The British radio monitors, based in the Admiralty's Room 40, were once again able to give timely warning of the impending major German deployment, and once again the Grand Fleet was on the move before the Germans set out.

Jellicoe led twenty-nine dreadnoughts and Beatty brought six battlecruisers, all with the usual escort of smaller ships, into the North Sea from the Scottish bases, and twenty-five submarines were deployed along the east English, Dutch and German coasts. Six light cruisers and nineteen destroyers from the Harwich Force were brought up, and the British were confident that their swift preparations would be rewarded with a major action. First blood went to the Germans when the *U52* sank the light cruiser *Nottingham* off the Farne Islands east of Northumberland with torpedoes. Jellicoe's abiding nightmare of a major submarine assault leading to disaster for his big ships led him to put about and steam north for two hours. He thought he had the time: the Admiralty had misread an intercepted German wireless message as meaning that the High Seas Fleet was much further east still than was the case. The message had come from a battleship left behind after being damaged by a British submarine torpedo.

Scheer was misled in turn by a Zeppelin pilot who mistook a section of the Harwich Force for ships of the line. Here at last apparently was the isolated section of the Grand Fleet he had prayed for. He abandoned the raid on Sunderland and turned south-east, a happy accident which saved the High Seas Fleet from potential disaster, not that either side knew it at the time. Had he stayed on his original course, Scheer would have been cut off by the Grand Fleet, now coming south again. As it was, the Germans, failing to find anything to attack, put about before reaching the minefields off the Humber and made for home.

A disappointed Jellicoe reversed course, and came under attack by German submarines on his way north. The light cruiser *Falmouth* was torpedoed and sank. The Harwich Force ships meanwhile chased the retreating German fleet with the idea of overtaking it and charging it from forward. The High Seas Fleet was, however, moving too swiftly and was too close to home. The British, although they did not know it at the time, had missed their last chance for a fleet action: the only later outing of the High Seas Fleet's battle line was a half-hearted sweep as far as the Dogger Bank on 18 October, abandoned because of bad weather and news of British torpedo damage inflicted on a light cruiser by a submarine.

These two ineffectual naval brushes had the most extraordinary and disproportionate consequences. The August outing led the Grand Fleet to abandon most of the North Sea for lack of destroyers; the October run led the High Seas Fleet as such to stay off the North Sea altogether for lack of submarines. In September the British Admiralty accepted Jellicoe's proposal, backed by a no less cautious Beatty, that the Grand Fleet should come no further south than the Tweed or further east than longitude 4° except in dire emergency unless 100 destroyers were available, which was inconceivable given the situation in the Channel. The Grand Fleet therefore stayed in port and ignored Scheer's aborted sortie on 18 October. Scheer for his part refused to contemplate taking the fleet out again unless he could deploy all available submarines for reconnaissance and torpedoing. Since the Germans resumed U-boat raids on merchant ships in British waters in September and these once again had priority, this demand too was impossible to fulfilment. For the lines of battle, the war was over.

Jutland was to stand as the last fleet action in the Nelsonian tradition of two lines of opposing broadsides, not only of the First World War but forever. The battleship had already been checkmated by the submarine. The great bulk of the vast opposing naval forces were to meet only once more, and that was after the Armistice which ended the war. Some sixty-five of the world's most powerful, most expensive and most lethal warships spent the rest of the war on the sidelines, their crews chafing in the German ports or enduring the bleak boredom of Scapa Flow, and an eerie quiet descended on the North Sea, which was abandoned to the fish and the invisible submarine. Scheer now admitted that he could see no other role for his magnificent capital ships except as mother-hens for the submarines as they left or returned to home waters. The submarine war now dominated German naval thinking; but the Kaiser delivered a personal dressing-down to Scheer when the Admiral sent out five dreadnoughts to back the recovery by destroyers of two U-boats in trouble off Jutland in November 1916. A British submarine damaged two battleships with torpedoes and only one of the U-boats was rescued.

If the North Sea was now an empty theatre of war, the same can hardly be said for the Channel, where a vital and rather more evenly matched struggle was going on between the dashing small ships of the Royal and Imperial navies. The Channel Fleet of nineteen British pre-dreadnoughts was in being at the onset of war but soon found itself with little enough to do. The German battle fleet was hardly likely to come so far south when it could so easily be cut off from home by the Grand Fleet. But there was a great deal of varied work for the Harwich Force and the Dover Patrol once the German advance through Belgium gave the Germans control of the forty miles of the Belgian coast and the naval bases of Ostend and Zeebrugge.

The Dover Patrol was the main force called into being at the beginning of the war especially to protect the Channel. Under the command of Vice-Admiral Bacon, it was usually made up of twenty-four destroyers, two light cruisers as flotilla leaders, eight 'P' (fast patrol) boats and a heterogeneous collection of trawlers, drifters and other conscripted small boats. There were also fourteen Royal Navy 'monitors', strange vessels with the elegant proportions of lidded bathtubs, an adaptation of the wallowing ironclad USS

Monitor developed by the North in the American Civil War, which gave its name to the type. Understandably perhaps, these vessels were not dignified with the appellation 'ship': no HMS prefix for them, just HM Monitor. Broad-beamed, flat-bottomed and slow, especially in high seas, the monitors carried a pair of 15-inch guns in a high turret and a dozen lesser pieces. They were designed as floating artillery batteries to bombard the German-held shore and, for a time, to take pot-shots at Zeppelins.

Harwich Force, led by Commodore Tyrwhitt throughout the war, played a subsidiary but important role in the Channel (it was often called northwards in support of the Grand Fleet, as we have seen). Tyrwhitt normally had five light cruisers (the Fifth Squadron) and some 36 destroyers (the Ninth and Tenth Flotillas) with four more light cruisers as flotilla leaders. Elements from both British light formations combined to patrol the eastern side of the Channel from the French port of Dunkirk.

The British Channel forces were always busy. They had to cover the ceaseless procession of troop transports taking men to the Western Front; they laid mines; fired on German shore installations in Belgium; covered the seaplane-carriers which launched air-raids on enemy positions; searched for German submarines and mines and patrolled. They always had to watch for German raiders who had the advantage much of the time because their role was almost exclusively offensive, while that of the superior British forces was mainly defensive, which meant they had to be much more versatile.

After Jutland, the Germans stepped up their activities in the Channel, the only naval area where it was now open to them to do so. The decision to keep the High Seas Fleet at home and to resume the submarine campaign in earnest released German destroyers to boost their Channel forces. Only a week after Jutland, a dozen destroyers briefly menaced Dunkirk before returning north. There were already about two dozen German ships based on the Belgian coast, and they had the initiative in choosing times and targets.

After the sweep by the High Seas Fleet in October, Captain Michelsen more than doubled the German destroyer forces in the Channel by flitting down the Dutch coast to Zeebrugge with twenty-four ships, unchallenged. Two nights later he took them out to raise havoc in the Channel, attacking net drifters guarding submarine defences, sinking a British destroyer and damaging other vessels. In the process they destroyed the forward half of the destroyer *Nubian* on their way home. The stern half was towed into port and joined to the forward half of her sister-ship *Zulu*, which had lost her stern to a mine. The unprecedented Lego-tactic worked, the graft took hold and HMS *Zubian* was born. But Michelsen's raid embarrassed the British Government and the Admiralty, which ordered Tyrwhitt to send heavy reinforcements to the Dover Patrol while more destroyers came down to Harwich from the Grand Fleet. One of Michelsen's flotillas returned to Wilhelmshaven and the Germans were rather more cautious for a while, without in any way reducing the embarrassment their continued existence represented for the Royal Navy.

Meanwhile, the German submarines were back in strength when raids against shipping in British waters were resumed. At the start of the war, the Germans had just one U-boat flotilla. In autumn 1916 when the raids resumed, there were four, plus the separate Flanders Flotilla, available for operations in the North Sea and Channel (the Baltic, the Black Sea and above all the Mediterranean had other U-boat flotillas, the last mentioned being particularly strong and effective). The resumption of raids was under prize rules, which meant challenging before shooting, something which could be done only on the surface, throwing away the one advantage the U-boat had. As the early U-boats had extremely noisy petrol engines which belched flames from their exhaust at night and they took several minutes to submerge (it could hardly be called 'diving' in the early days) there were enough risks already, together with armed merchantment and disguised Q-ships to worry about. Nonetheless, even following prize rules, the German boats were doing better and better and were now capable of staying at sea for weeks at a time. They were becoming larger, quieter, faster and more versatile as the relevant technology progressed by forced growth under pressure of war. In June 1916, the U-boats sank 109,000 tons of shipping in all sea areas. In October, the month raids were stepped up under prize rules, they sank 337,000 tons, a rate maintained until February 1917, when Scheer once again ordered unrestricted raids without warning in British waters, as had prevailed from February to September 1915. It was a rather more serious threat than two years earlier because Germany had distinctly better and more numerous boats and because she was now committed to the U-boat campaign as never before.

The idea was to knock Britain out of the war in five months as a way out of the deadlock on the Western Front. The method was starvation, the same weapon the British were using against Germany with their mines and blockade. The distinction between the two methods was one of style, not motive. The Germans knew the Americans were totally opposed to submarine warfare and that they were therefore risking United States intervention in the war on the Allied side, but they calculated, not unreasonably, that they could cripple Britain before the Americans could affect the course of the war. There are no prizes for being nearly right in war, any more than there are for being runner-up in a boxing match. The German mistake was not to go wholeheartedly for the U-boat option soon enough, which prevented them from developing it to the extent necessary for victory. At the root of this was the vacillation of the Kaiser, exacerbated by a division of opinion in his navy at top level. Now, however, the U-boats gained top priority in high-grade steel, shipyard space, finance, manpower and training. All other naval operations were subordinated to the campaign. The High Seas Fleet was used to provide manpower and officers, to lead the U-boats in and out, to attack British anti-submarine operations, and to join in the unrestricted attack on merchant shipping. The Germans also stepped up their efforts in the new art of minesweeping to counter increased British minelaying (and their adoption of more effective horned mines). On 1 February 1917 the Germans had 106 U-boats with which to open the new all-out campaign.

The first three months seemed to prove the Germans right. The unleashed, unconstrained U-boats sank more than 500,000 tons in February, nearly 600,000 in March and almost 900,000 in April, the peak month of the whole war. Well over half these catastrophic losses were incurred in British home waters, the Western Approaches, the Channel and the North Sea. The campaign was to reduce Britain to six weeks' rations for her population of forty million and the Admiralty seemed incapable of finding an answer. In these worst three months for the British, Germany lost just nine U-boats. But on 2 April 1917 America at last declared war on Germany, a fact which was to outweigh by far the imminent German defeat of Russia, effected by August and formalised by treaty in December. Despite the overwhelming success of the opening phase of the renewed U-boat campaign, Scheer had an attack of doubt soon enough. Germany had built more than forty new boats in the intervening months but was unlikely to be able to maintain a fast enough construction programme to keep up with the inevitable British counter-measures. The Americans had forced Germany to restrain the U-boats in April 1916; now they were in the war anyway and Germany had lost a year. The promise to the Kaiser to bring Britain to her knees by the end of June 1917 could not be delivered. Already in May Allied shipping losses were reduced by one-third compared with the peak month of April. Lloyd George, now Prime Minister, had at last forced the Admiralty to introduce the convoy system in that same catastrophic month. Losses fell below 200,000 tons in August and went on falling, to a 'tolerable' level of 1.25 per cent of sailings. Hundreds of patrol craft were deployed by the British; the hydrophone and the depth-charge were further developed.

Meanwhile the Germans found they could not replace their lost U-boats fast enough once the kill-rate went up. More important, there was a growing shortage of crews and officers. They had 138 U-boats in commission in all areas at the beginning of 1917, during which they lost seventy-two but built 103. In 1918 Germany built eighty-one and lost eighty-one, ending the war with 169 boats in service. Total U-boat losses throughout the war amounted to 199. The British made a colossal effort to close the Dover passage to U-boats, using mines, nets, 100-ship patrols, searchlights and aircraft. The Germans were driven to favour the long haul round Scotland. Then the incredible total of 70,000 mines were laid between the Orkneys and Norway in March 1918, a feat out of all proportion to the catch of just six U-boats. But the convoy was the tactic which eventually frustrated the U-boat. Allied admirals were slow to see it, partly because their instinct led them to measure success by the number of enemy boats destroyed and partly because other methods such as patrols, aircraft and mines actually caught more U-boats than convoy escorts. It was the preventive effect of the convoy system that explained much of its success, and there was, and still is, no method of measuring non-events or recording the number of times U-boats turned away from the protective screen. The important factor in the closing stages of the war was the number of Allied ships that were *not* sunk, not the number of U-boats that *were* sunk, even though the latter figure rose.

Yet the Germans' last throw at sea, too little and too late though it proved to be, accounted for nearly 2.5m tons of British shipping in six months and another 1.5m tons of Allied and neutral shipping. Disaster for Britain was narrowly averted by rationing, indemnification of neutrals to cover their shipping losses in trade with Britain and an enormously accelerated shipbuilding programme, augmented by America, the world's richest nation and its leading industrial power. The German strategy therefore was after all defeated by the entry of America into the war and her own incapacity to sustain the

U-boat offensive and expand it. The American contribution was of course to be decisive; it was too late to stop it and Germany's resources were too thin to force peace on Britain before it began to tell. The only hope left to the Germans after July 1917 was that Britain and France would tire of the war sufficiently to enable Germany to conclude a reasonable peace. And by the end of 1917 the Germans realised that even the U-boat campaign was not going to succeed. The convoy system covered all shipping in British waters and the North Atlantic (already U-boats had shown that they could reach the American coast). Fortunately for the Allies, the greatest challenge to the convoy, the U-boat pack, was not developed in this war but only in the next. A U-boat campaign did, however, thrive in the less important theatre of the Mediterranean, assisted by the continued lack of a convoy system there and the failure of the British, French and Italian navies to co-ordinate tactics. The attempt to construct a barrage across the Straits of Otranto to stop German U-boats using the Adriatic bases of Austria was a costly and abject failure. The Austrians destroyed it, and won the ensuing naval action.

Once the last strategic effort at sea, the submarine campaign against Britain, had failed, the Germans drove their war-weary troops into a desperate great offensive on the Western Front. Taking advantage of a 30 per cent reinforcement of troops made available by the collapse of Russia, General Ludendorff launched the heaviest and most menacing German assault since the very beginning of the war on the Somme. At home, Germany was in desperate straits. Acute shortages had accentuated the general war-weariness and eroded the national will to fight. Time was running out. Scheer's High Seas Fleet came under pressure to abandon its inactivity, which contrasted so markedly with the herculean effort of the German army, and to make a real contribution at last in support of the troops. Scheer decided upon a new variation of the old idea of a sortie to destroy a detached part of the Grand Fleet.

In the later part of 1917 the Germans had tentatively essayed to interfere with the convoy traffic between Britain and Scandinavia. In December they sank an entire convoy and one of its escorts, a stroke which led the British to provide a battle squadron to protect each convoy. The Germans knew of this new system, and Scheer saw his best opportunity so far to achieve the elusive aim of destroying a significant part of the Grand Fleet by overwhelming one of these squadrons of battleships with the whole High Seas Fleet, reducing the disparity significantly. Beatty, in command of the Grand Fleet since 1917, had thirty-four dreadnoughts and nine battlecruisers to Scheer's nineteen and five; a British battle squadron usually consisted of eight capital ships.

It was a sound enough scheme. This time Scheer gave maximum attention to secrecy, passing off his assembly of the fleet in the Schilling Roads as an exercise, and imposing radio silence. The only drawback was that there was no convoy at sea that day, as German intelligence should have known. But total secrecy was achieved, for once, on 23 April, the day of the last sortie of the High Seas Fleet. Scheer's plan was to send Hipper ahead with the battlecruisers to locate the expected convoy, attack it, and lead the escorting battleships on to the guns of the German battle line. Unreported by a British submarine which assumed they were friendly, the German ships sailed north-westward for twenty-four hours, the furthest they had ever been from their bases. Then the battlecruiser *Moltke* had a serious mechanical breakdown. She fell back on the battleships and her condition led Scheer to decide to return home with them, leaving Hipper at sea to destroy the awaited convoy. He found nothing because there was nothing to be found, having searched the area as far north as south-west Norwegian waters before turning back and catching up with the homeward-bound battleships.

Meanwhile, the *Moltke*'s radio calls for help had alerted the British. This is more than can be said for one of their submarines, which saw the German ships *twice* during their sortie and failed to report. Beatty took the entire available strength of the Grand Fleet out of the Firth of Forth, now its main base, and headed east with thirty-one battleships and four battlecruisers, with the usual escorts. It was their last excursion of the war, too, and they were twelve hours too late. The last seaborne action of the war of any significance, the British raid on the German base at Zeebrugge, had taken place on the night of 22 to 23 April. No contact at all occurred between the two fleets and the Germans got home on the evening of the 25th. An isolated British submarine torpedoed the limping *Moltke*, but she still managed to get home some hours later. Beatty turned back, a bitterly disappointed man, and also very concerned that the Germans this time had managed to get out without his knowledge.

The North Sea stalemate lasted until the end of the war, which the British decision to maintain maritime supremacy at all costs, including battle honours, had done so much to win. The greatest contribution of the Silent Service was its decision not to sacrifice its power, itself a major sacrifice for the heirs of Nelson. The upstart German navy now proceeded to succumb to its own frustration. For the High Seas Fleet, the real ordeal was about to begin.

PART III

The Grand Scuttle – The Salvage of Honour

With ships the sea was sprinkled far and nigh.

– William Wordsworth, *Miscellaneous Sonnets*

The deed is all, not the glory.

– J. W. von Goethe, *Faust*

CHAPTER FIVE

Mutiny

THE FATE OF the German navy was naturally a major issue in the negotiations and discussions which eventually led to peace. But it was not only a bone of contention between the victors and the vanquished; it also divided the Allied camp vertically and horizontally, putting the French and the British at odds with the Americans, and the Allied Naval Council of officers at odds with the Supreme War Council of their political masters. After the German Government offered an armistice on 5 October 1918 on the basis of the Fourteen Points drawn up by American President Wilson, one of the first responses by the Allies and the Americans was to press for an end to the German submarine campaign against merchant shipping. Following Berlin's agreement on 20 October, the raids were halted.

The German Admiralty was as aware as anyone else in Germany that the withdrawal of the army in the west after the Allied counter -offensive at the end of July 1918 meant an imminent end to the war. But unlike the Generals, the Admirals had at their disposal very powerful surface forces which had been frustrated but not defeated and remained virtually intact. They therefore drew up plans for a great last throw which was intended to influence the peace negotiations decisively in Germany's favour. The idea was not new. The Dutch had conceived and daringly carried out a similar stroke at the end of the second Anglo-Dutch War in the seventeenth century with their dashing raid on the English navy in the Medway, which had the effect of strengthening their bargaining position at the peace talks at Breda. But the German plan was much bigger in scale, large enough, at any rate in theory, to change the course of history. The idea was simple, bold, risky but far from desperate. Light cruisers and destroyers would launch a series of raids from the Belgian ports on the Thames Estuary on a scale sufficient to draw the Grand Fleet southwards from its Scottish bases. New minefields would be laid across its path and submarines would be lying in wait for it. The High Seas Fleet, now under Admiral Hipper, would put to sea with every available capital and supporting ship to assault what it was hoped would be an already diminished and disrupted Grand Fleet at a time and a place of the Germans' own choosing. This plan, pursued to the utmost, could have been a belated but spectacular vindication of Tirpitz and his risk theory, weakening the British by reducing their naval supremacy and strengthening the German negotiating position as a result. The beauty of this stratagem was that it did not matter whether the Germans won or lost, provided only that the High Seas Fleet inflicted severe damage on the Grand Fleet, on paper a fairly safe bet. The German plan was, therefore, far from being a madcap final fling or a last, desperate gamble by the Prussian Officer Corps to salvage its honour; it was a calculated risk, strategically and tactically sound and of a strictly military nature which posed an unprecedented threat to Britain in its hour of victory. The German Admirals were not about to throw away their fleet: on the contrary, it would at last earn its keep after years of frustration just when Germany needed it most.

Some of the submarines which had been withdrawn from raiding merchant shipping were therefore sent to positions off the east coast of Scotland to lie in wait near the Grand Fleet's bases. Some of them were detected by the British, without alarm or the slightest inkling of the German plan. The British Admiralty in fact could not have been more wrong about the German Admirals' intentions.

The British were correct in calculating that the Germans would see the High Seas Fleet as an important bargaining counter at the peace talks, but they went on to conclude that this meant the German fleet would not venture out again so that its value would be preserved intact. The Grand Fleet stayed in harbour and did nothing about its low state of readiness to put to sea.

The High Seas Fleet set sail on 29 October for its usual assembly point at the Schillig Roads in slightly greater strength than it mustered even at Jutland. There were twenty-two capital ships, a dozen light cruisers and some seventy destroyers in seven flotillas. At this point, however, the dangerous plan of the German Admirals – dangerous to the Allies, dangerous for the German fleet – collapsed. They had reckoned without the low morale, war-weariness and general discontent of their own crews. Mutiny broke out on some of the capital ships and soon became general, despite firm initial counter-measures by Hipper which briefly appeared to have succeeded. When parts of the crews of two battleships refused orders, a submarine and five destroyers were cleared for action, their torpedo tubes loaded and lined up on the looming sides of the two capital ships at point-blank range. Some of the guns on the battleships were seen to move, though it was said later that some of their gun-crews were merely using the optical equipment in the turrets to get a better view of the proceedings, a singularly stupid act under the circumstances. The mutineers surrendered to the ultimatum and were arrested, to be taken to prison in Wilhelmshaven. Hipper was forced to abandon the operation with unrest simmering throughout the fleet, which began to disperse back to its bases. When the Third Squadron of battleships returned to Kiel with a number of agitators in irons, a general mutiny broke out on shore in response.

By the end of the first week in November, Workers' and Soldiers' Councils on the recent Soviet model had been set up in Kiel and Wilhelmshaven (where the Republic of Oldenburg was briefly created) and red flags were hoisted at many points along the North Sea and Baltic coasts, including Hamburg. It seemed that the crucial role played by sailors in the Russian Revolution was about to be repeated by their German counterparts. They certainly thought so as they set about administering an unparalleled shock to a nation second to none in its passion for law and order. The naval mutiny matched and surpassed the recent action of army units on the Western Front, where whole divisions had refused to fight.

Chaos broke out in many parts of Germany, notably in Berlin and Munich, where another Soviet-style republic was set up. The main naval bases were reduced to a shambles and remained in that condition even after the government promised, in a decree of 4 November, that the navy would not be sacrificed. The High Seas Fleet, still physically intact, was effectively swept off the board. With the heart gone out of its crews, the actual value of the world's most modern fleet was no more than what it might fetch as scrap. One group of officers, moved to shame and desperation by the mutiny, stole aboard the submarine *U18* and took her out to sea at the beginning of November 1918. They planned a suicide raid on those ships of the Grand Fleet still in Scapa Flow, four years almost to the day after the only other attempt to raid the anchorage during the First World War. The submarine, however, was sighted by an Admiralty trawler off the Scottish mainland and rammed by a destroyer.

The armistice negotiations continued against this volatile background. On 9 November the Kaiser abdicated and fled to the Netherlands, and within two days the Armistice was signed at Compiègne in northern France. The document to which the exhausted negotiators put their signatures was, in effect, a draft of the Peace Treaty which was eventually signed at Versailles on 28 June 1919, although the negotiations and Allied demands led to a long series of additions and amendments. The original Armistice period was thirty-six days but it was repeatedly extended as the peace talks dragged on.

Article XXI provided for the surrender by Germany of its entire submarine fleet, a provision on which the five allied and associated powers – Britain, France, Italy, Japan and the United States – found no difficulty in concurring. Eventually getting on for 200 boats were handed over, most of them at Harwich within two weeks of the Armistice. These were later shared out among the victors and most of them were scrapped. But the three principal victorious powers, Britain, the United States and France, were at odds on what to do about Germany's surface vessels during the Armistice and even more so in deciding on their ultimate fate once the peace treaty had been signed. The Allied Naval Council was less divided. The British, French and American representatives determined that the entire navy should be disarmed and confined to specific German ports under Allied supervision and also that Germany should hand over a total of seventy-four ships, ten named battleships, six named battlecruisers, eight named light cruisers and fifty of the most modern torpedoboat-destroyers, unnamed but to be specified later. The only point of disagreement was the basis on which they were to be handed over. The British and French Admiralties wanted them all to be surrendered, while the Americans wanted the ten battleships to be interned in a neutral port. This difference of opinion in the Naval Council became irrelevant when the Supreme War Council, consisting of Lloyd George and Clemenceau, the British and French Prime Ministers, and Colonel House, representing President Wilson, over-ruled the naval men by insisting that the seventy-four chosen ships should be interned and their final destiny decided at the Peace Conference. The dispute between the Admirals and the politicians lasted a week and the naval men made no secret of their reluctance in bowing to the will of the statesmen. They accepted internment, but insisted that the ships should never be given back to Germany and placed their disagreement on record.

The provision relating to the seventy-four ships to be handed over went into the Armistice terms as Article XXIII, which stated:

> The German surface warships, which shall be designated by the Allies and the United States of America, shall forthwith be dismantled and thereafter interned in neutral ports, *or, failing them, Allied ports*, to be designated by the Allies and the United States of America, only care and maintenance parties being left on board. The vessels designated by the Allies are . . .

The ships were named, but as there were several changes for various reasons they need not be listed here (the ships which were actually interned appear in the Appendix). The words italicised above were inserted in the original draft by the naval men. They took the view that the politicians might insist on internment, but the practical difficulties of internment in neutral ports would be insuperable, because the degree of supervision of the confined ships necessary to prevent sabotage or a dash for freedom would come unacceptably close to infringing the sovereignty of the neutral countries concerned. The only neutral country which appears to have been formally approached by the Allies with an inquiry about provision of internment ports was Spain, which refused. The Germans received negative replies from all other likely neutrals when they asked whether the Allies had sought facilities from them. Thus on the very day the Armistice came into force, the Allied Naval Council took the decision that the place of internment could only be Scapa Flow, the principal anchorage of the Grand Fleet – the only force capable of making internment stick under any conditions. The decision in principle was confirmed at a second meeting on 13 November, when responsibility for organising the internment was handed over to the Commander-in-Chief, Grand Fleet, Admiral Sir David Beatty (created an Earl in 1919). He was given effectively a free hand.

One of the main reasons for the naval commanders' objection to internment rather than surrender was their early recognition of the possibility that the Germans might scuttle their ships if the eventual Peace Treaty required a permanent handover of the interned fleet. Internment meant that German skeleton crews would remain in possession of the ships and that Allied guards could not be placed aboard because the interned vessels would, in terms of international law, still be the property of Germany. No matter how vigilant seaborne patrols might be in keeping watch, any naval officer knew that once the seacocks of an interned ship had been opened and it began to sink, it was highly unlikely that anything could be done to save it by boarding and seizure. Thought does not appear to have been given to the idea of interning the ships in a carefully chosen anchorage in relatively shallow water – the Allies controlled large areas of the almost tide-free Mediterranean – where an attempt to scuttle would be pointless because the ships would touch bottom and at worst capsize for lack of deep water under their keels. Although Scapa Flow varies in depth, none of it is shallow – hence its value as an anchorage. The possibility of a decision to scuttle was anticipated by Article XXXI of the Armistice, which stated: 'No destruction of ships or of materials is to be permitted before evacuation, delivery or restoration.'

The Germans, however, began to think in terms of scuttling very soon after learning the terms of the Armistice as it affected their navy, even though the decision to intern the ships at Scapa Flow was not officially passed to them until well after it was a fait accompli. The German argument that their ships should go to a neutral port because they had not been defeated fell on deaf ears before and after the handover.

As if to concentrate the German Admirals' minds, the Allies supplemented the naval conditions of the Armistice by informing the Reich's delegation that they would seize Heligoland if the ships to be interned were not ready to sail on 18 November, seven days after the Armistice began. The threat, made on 11 November, would be carried out if the mutinous state of the crews held up the transfer. It proved effective. Notices drawn up by the Workers' and Soldiers' Councils went up in the ports calling for a maximum effort for the sake of the Fatherland to get the ships ready in time. 'If we fulfil the conditions, the ships will then return when peace is concluded; if we do not fulfil them, then the Englishman will come, take away our entire fleet forever and bombard Wilhelmshaven.'

Preparations for the internment went ahead at a feverish rate on both sides of the North Sea. During the night of 12 to 13 November, a radio message from Beatty reached the High Seas Fleet command requesting the despatch of a flag officer in a light cruiser to complete arrangements for the fulfilment of the naval conditions. Rear-Admiral Hugo Meurer left Wilhelmshaven aboard the *Königsberg* on the afternoon of the 13th to carry out this galling and difficult mission. In tranquil weather the cruiser followed a circuitous route of 430 miles to avoid minefields on her way to the designated rendezvous point fifty miles due east of the Isle of May and the Firth of Forth. As she passed the coast of Jutland, a neutral Danish cruiser fired a salute. The *Königsberg* had been disarmed and could not reply in kind so she signalled her apologies and thanks. In the North Sea on the 14th, the cruiser's War Diary records, she met a German submarine, the *U67*, which asked her for details of what had happened in Germany.

The British mounted 'Operation Z1' to receive her. The *Königsberg* was met by the Royal Navy's Sixth Light Cruiser Squadron of five ships led by HMS *Cardiff* and escorted by ten destroyers which conducted her into the Firth of Forth on the evening of the 15th. Meurer was asked aboard Beatty's flagship, the *Queen Elizabeth*, the same evening. With him went three staff officers and his flag lieutenant. Beatty refused point blank to receive the three representatives sent by the Workers' and Soldiers' Councils, saying that he could not receive a delegation from a government which the British Government had not recognised. The three men thereupon gave Meurer a mandate to negotiate and sign on their behalf. The British C-in-C told Meurer that he assumed the German Admiral could speak for the whole German navy. Meurer replied that he had come as the representative of the C-in-C of the High Seas Fleet to explain the difficulties being experienced in meeting the Armistice conditions and to learn Beatty's wishes. Beatty said he was speaking for all Allied naval forces, and Meurer was obliged to send a telegram to Germany asking for plenipotentiary powers. The Germans were handed papers detailing the procedure for the transfer of the ships to be interned and were given until 9.30 the next morning to reply. The three sessions of talks on the 17th soon showed that the documents were not discussion papers but orders. The only measurable concession related to the disarming of the guns aboard the capital ships. The British wanted the breech-blocks removed, but when Meurer pointed out that this would entail the opening of the gun turrets, Beatty agreed that there was not enough time for such drastic measures and accepted instead the removal of the firing mechanisms of the guns. After the third session, a telegram arrived from the German Armistice Commission giving Meurer full powers to speak and sign for the entire navy. At the fourth and final session late that evening, Beatty read out the conditions for the transfer of the forfeited submarines and the surface ships to be interned and Meurer signed them. He got back to the *Königsberg* at 3.30 a.m. on the 17th and the ship sailed for Germany under escort one hour later.

Meurer found the atmosphere at the talks distinctly frosty. 'The sessions were led by Admiral Beatty in a thoroughly formal and businesslike manner, but clearly without the slightest personal compromise. All questions were decided in his favour without discussion . . .' All appeals for consideration of the sense of honour of the crews by avoiding humiliating procedures 'struck a cool refusal'. No sympathy was shown for the problems of the German navy arising from 'the miserable situation in Germany' and no concessions made, Meurer wrote in his report. He noted, no doubt with nostalgia and envy, the discipline, order and bearing of Royal Navy personnel, and recognised that there was not the slightest chance of a realisation of the German revolutionaries' dream of a sympathetic peace offensive breaking out in the British Fleet. Meurer clearly gained the impression, borne out later by Admiral Reuter's experience, that magnanimity in victory was not Beatty's strong suit. Having radioed the main points of the British orders for the handover, Meurer took the papers back to Wilhelmshaven for execution. They went into the most minute details and were obviously the product of intense and rapid staff work by the British, who were prepared to leave nothing to chance. There was however no clue to the final destination of the ships to be interned.

In a message to the Admiralty of 17 November, Beatty took a rather different, not to say smug, view of how the talks with Meurer had gone. He and his fellow-Admirals had got the impression that 'the representatives of the High Seas Fleet were prepared to agree to any demands we made and that they preferred their fleet being moved to a safe place to the alternative of it being controlled by the Workmen's [sic] and Soldiers' Council.' He said that the German officers were 'pleased' by his refusal to have anything to do with the three delegates of the Republic of Oldenburg.

Tangential though it may be in the context of an account of the death of the Kaiser's navy, the irony of Beatty's impending triumph and the character of the man who was about to take delivery of the biggest stroke of luck to come the way of an English navy since the Spanish Armada was smashed by storms in 1588 cannot be passed over. One of his captains said of him: 'Beatty was born with such an outsize silver spoon in his mouth that I'm surprised it didn't choke him.' He married an American millionaire's daughter, he won the DSO as a lieutenant, he was a commander at twenty-seven, a captain at twenty-nine and attained flag rank at thirty-eight in 1910, Britain's youngest Admiral since Nelson. Churchill in *The World Crisis* quotes another Admiral, Pakenham, as saying of Beatty in 1915: 'Nelson has come again.' The photographs show a handsome figure, cap jauntily tilted over the eye. Dashing, gifted and forthright, Beatty manages to seem arrogant even in death: his stark and imposing tombstone in the crypt of St Paul's Cathedral in London bears in gold letters the one word, 'Beatty'.

Yet the chance to be another Nelson was denied him – the Germans failed to provide a Trafalgar for him to win. He did his duty as head of the Grand Fleet from the end of 1916 by bottling up the German navy after Jutland and preserving the Royal Navy's supremacy, a vital strategic contribution but hardly glamorous or Nelsonian. If Hipper's abortive last foray had come off and Beatty had been given the chance to smash the enemy at sea once and for all, he would surely have been in his element. The most spectacular moment in his career, when the disarmed bulk of the High Seas Fleet accepted defeat without a shot being fired and presented itself for internment under his tutelage, was the hollow climax to an extraordinary career which he completed as First Sea Lord after the war. Even his official correspondence with the Admiralty at the time of the internment reveals a man trying to make the best of what he clearly regards as a bad job. It is thus possible to feel sorry for the still remarkably young full Admiral of forty-eight in the hour of his triumph at 8 a.m. on 21 November 1918, when the High Seas Fleet kept its rendezvous with him, in accordance with the terms he gave Meurer.

Nobody in Wilhelmshaven, where the High Seas Fleet command was based, believed that the ships could possibly be ready in time to meet the British deadline for delivery. The only people capable of organising the necessary preparations were of course the officers, whose authority was continually flouted by the ratings and undermined or countermanded by the Workers' and Soldiers' Councils, until the latter resolved to go along with internment. Word of the Allied threat to seize Heligoland spread round the fleet on 14 November. The beleaguered, desperate and largely rejected officers did their best to adapt to the chaos and get the work of disarming and supply completed, many of them showing remarkable stoicism, ingenuity and devotion to their melancholy duty, giving the lie to the common belief abroad that the Prussian Officer Corps tradition, as strong, though different, in the navy as in the army, could produce only automatons. The accumulation of setbacks they had to contend with was uniquely daunting for men who had lived under Standing Orders which forbade them to smoke in public, dance the tango or talk politics with the ratings. Now they needed special permission from the Soldiers' Councils to wear their epaulettes, swords and other trappings of their rank; they were constantly insulted; 'through the revolution all restraints of obedience and discipline were completely dissolved in the Navy', wrote Admiral von Reuter in his official report on the internment.

A fleet already weakened by an influenza epidemic in August and constant calls on its manpower for the submarines in the closing stages of the war was further reduced by wholesale desertions, drunkenness and general indiscipline. Some men brought women aboard for orgies and much of the fleet succumbed to an irrational passion for dancing. Somehow volunteers went ahead with the work in a travesty of a carnival atmosphere to the sound of music and singing, cigarettes hanging from their lips.

By no means all the rank and file enjoyed the anarchy. Commander Hermann Cordes, who was to be leader of the torpedo-boat-destroyers at Scapa Flow, noted in his report on the events surrounding the internment that many ratings exhibited shame over their own and their comrades' indiscipline and untidiness. Morale on the smaller warships never sank as low as it did on the capital ships either in the revolutionary period or during internment. Officers and men knew each other better and the Soldiers' Councils on the destroyers were more moderate. Once a few highly unpopular officers were sent packing, a workable relationship between the remainder and the Councils was established quite quickly and effectively.

Six decades after the shattering events of the end of the war, Seaman Werner Braunsberger of the battleship *Kaiser* recalled the shaming of the fleet with vivid bitterness. 'We were ordered to tear the letters "S.M." [His Majesty's] out of our cap-bands so that only the letter "S" [Ship] and the name of the ship could be read. The black, white and red [the Imperial tricolour] of the cockade on our caps had to be painted red. The officers too had to take off their badges of rank. There were cases where officers who wanted to resist this order had their badges of rank torn off by the dehumanised revolutionaries. So much for the beginning of the end of the Imperial German Navy.'

Haphazardly but with remarkable speed under the circumstances, the ships were made ready for their voyage into internment. Gunparts, gunnery control equipment, rangefinding apparatus, ammunition, small arms, even signal flares piled up in disorderly heaps ashore. The most secret equipment which had helped the German ships attain unsurpassed standards of gunnery was flung carelessly on to the cobbles and concrete of the quays. The disorderly crews showed inordinate enthusiasm for the work of destruction which reduced their ships to impotence. It proved rather more difficult to find men to help with the more awkward disarming tasks such as the removal of the firing mechanisms from the breech-blocks. The revolutionary leadership of the Republic of Oldenburg and the twenty-first committee of the Workers' and Soldiers' Council which presided over the shambles in Wilhelmshaven under the guidance of the President of the Republic, Leading Stoker Kuhnt, helped as little as possible. Much attention was devoted to the attachment of pine branches to the masts of some ships as peace symbols. The revolutionaries would not open the supply offices and depots they had seized; they made difficulties about releasing food stores to the men they purported to lead between orgies of gluttony and drunkenness in the officers' messes. False rumours that the British sailors were about to do the same thing, or had already begun, probably encouraged some of the demilitarised rabble to press ahead with the preparations for internment with unreliable and erratic gusto. Only the passive majority in the end made it possible to complete the task, somehow.

Meanwhile as the deadline drew closer, Hipper turned his mind to the task of finding a flag officer for the unspeakable assignment of taking the ships into internment. His original intention had been to give the job to Meurer, as the man who had made the first direct contact with the British. But Meurer had been obliged to spend more time with the British than the Germans expected and fog had further delayed his return to Wilhelmshaven. With time running out, Hipper fastened upon Rear-Admiral Ludwig von Reuter as the man to lead the voyage into the unknown. Reuter considered his options and, after due deliberation, decided to take on the responsibility. Meurer on his return was appointed to the German Armistice Commission in Wilhelmshaven and Reuter took command of the 'transfer formation' at noon on 18 November. What kind of a man was he, this Admiral suddenly charged with one of the strangest commissions in the history of naval warfare?

Hans Hermann Ludwig von Reuter was the fifth child and third son of an army colonel who died while commanding a regiment in the Franco-Prussian War of 1870, the year after Ludwig was born. It was a thoroughly military family: both his elder brothers became Generals and eventually all three of his sons became officers, two in the army and one in the navy. He was brought up in Coburg until his mother entered him as a cadet in the Imperial Navy in 1885 when he was 16. He became a midshipman a year later and sub-lieutenant in 1888. Reuter passed a seaman-officer's promotion examination at the end of 1889, but promotion was slow in the small, pre-Tirpitz navy and he became a lieutenant only in 1894. Four years later he attained the rank of *Kapitänleutnant*, equivalent in the Royal Navy to Lieutenant-Commander, the standard rank of a submarine captain in each navy. Reuter commanded the gunboat *Loreley* in the Aegean in 1901–2. Two years later he was promoted to lieutenant-commander (*Korvettenkapitän*) and in October 1908 he moved to the Imperial Navy Office in Berlin as a staff officer with the rank of junior captain (*Fregattenkapitän*), for which there is no equivalent in the British or American navies. As the German navy expanded at astonishing speed, so promotions were accelerated and it took Reuter only eighteen months to become captain (*Kapitän-zur-See*). In June 1910 he went to sea again as the captain of the heavy cruiser *Yorck*. This ship was named after a Prussian army commander of the early nineteenth century, and, in turn, Reuter gave the name to his eldest son.

In October 1912 Reuter was back on shore, this time as head of the central section of the administration of the naval docks at Wilhelmshaven. Two months after the outbreak of war he was given command of the battlecruiser *Derfflinger*, the navy's newest capital ship at the time.

Though rather past the age normally associated with dash and daring (at forty-five he was two years older than Beatty and it is difficult to regard men of that sort of age as promising), Reuter proceeded to have what the military calls 'a good war'. The *Derfflinger* took part in the shelling of the Yorkshire resort of Scarborough in December 1914 and in the Battle of the Dogger Bank the following month when the German battlecruisers under Hipper clashed with the British battlecruisers under Beatty. For his part in the action, Reuter received the Iron Cross First Class. He left the *Derfflinger* after exactly a year to become Commodore of the Second Reconnaissance Group of older-type light cruisers with the substantive rank of captain, a post he held for two years. He distinguished himself again in the Battle of Jutland when he took on a greatly superior British cruiser squadron. He lost one of his ships, but damaged the *Southampton*, the enemy squadron's flagship, by gunfire controlled from his own flagship, the *Stettin*, and won two more high decorations.

In November 1916 he was promoted to Rear-Admiral (*Konteradmiral*) and a year later became flag officer commanding the Fourth Reconnaissance Group of six modern light cruisers (flagship *Königsberg*, which took Meurer to Rosyth to meet Beatty). One month after taking command and eight days before his promotion to Rear-Admiral, Reuter took part in what was to be his last sea battle. Once again he acquitted himself exceptionally well.

On 17 November 1917 he was in the North Sea with four of his cruisers and a flotilla of ten destroyers, covering a large minesweeping operation. Suddenly the light German force came under attack from an overwhelmingly superior British one, including battlecruisers, battleships, cruisers and destroyers, more than two dozen ships. The nearest German capital ships, two battleships, were two hours away. Reuter fought a brilliant rearguard action with the loss of only one trawler. As he raced for the protection of his own battleships, he had the presence of mind to steer a course designed to lure the British ships into a German minefield, but the British decided to break off the action and abandon the chase just in time to avoid the trap. The *Königsberg* took a direct hit from a British 15-inch shell which turned out to be a dud. He was awarded one of the highest German decorations for his gallantry, and the unexploded shell stood for many years on a wooden pedestal in the family home.

The son recalls the father as a thoroughly 'Prussian' (his word) figure with high conceptions of honour and duty, yet endowed with a sense of humour and an independent spirit. An avowed monarchist, Reuter was sometimes fiercely critical of Kaiser Wilhelm II, the 'All-highest' he loyally served, who had the grace to remember him on his seventieth birthday, sending him a telegram from his Dutch exile in 1939. Reuter's personal hero among the Hohenzollerns was King Friedrich Wilhelm I, father of Frederick the Great. He enjoyed good company and had the gift of commanding attention as well as being a good listener. This model of a professional officer and gentleman was always concerned for the welfare of his subordinates and brought off the rare feat of being respected and popular simultaneously. He also had an eye for the ladies (but only an eye – he was apparently always correct), according to his son, who went on to say rather touchingly to the author that he realised his description of his father verged on the lyrical, but 'I have seriously considered telling you of some of his faults as well, but try as I might I cannot remember a decisive weakness.' It is a wise man who knows his own father, and the outside observer can detect in the Admiral's published memoir of the scuttling and his private report on it to his superiors a certain naivety and a tendency toward that particular brand of legalistic self-justification which is one of the curses of the German nation. But there can be no doubt that Ludwig von Reuter emerges clearly as a worthy hero, of a gratifyingly old-fashioned kind, in a most extraordinary drama, a man of action with a capacity, rare enough among such people, for reflection and even doubt, yet in the end decisive and responsible.

And it was indeed after due reflection that Reuter decided to undertake Hipper's commission (it was a request, not an order) to lead the main body of the High Seas Fleet into internment. His book, *Scapa Flow: das Grab der deutschen Flotte* (Grave of the German Fleet), makes the nature of his reflections clear.

The Admiral decided, as a servant of the German state, such as it was at the time, that the transfer was in his country's interests and therefore that his honour as an officer would not be compromised by his acceptance of such a personally repugnant task. On the contrary: the alternative, the threatened seizure of Heligoland by the Allies, would mean a total blockade of the German coast, which would be a new disaster to add to all the others. Either way, the Allies would take control of the High Seas Fleet. But if the Armistice were adhered to the interned ships would remain German property while a final peace was being negotiated, and the hope would remain alive that some at least of the ships would come home. The mutinous condition of the crews made any attempt to frustrate the handover futile in any case, and the Allies would have been able to walk in and help themselves. The state of the crews precluded sinking the ships (the revolutionaries were in favour of internment anyway) because the officers would have been unable to organise it. The ships had to go voluntarily to deny the Allies the opportunity of seizing them, using the mutiny as their excuse.

In his rather lengthier report to Berlin after the scuttling, Reuter wrote that his guiding principle was to save what could be saved. 'Personal feelings had to step to the rear. The German nation had unmanned itself of its own accord, the German forces were in no condition to resist . . .' If Germany broke the Armistice conditions, the enemy would have a free hand and there would be no limit to the ensuing disaster. The navy could not precipitate this. Reuter strongly denied that the officer-corps was guilty of a sellout in leading the ships into internment.

Before the last voyage he thought he would be taking the fleet to the Firth of Forth to allow the Allies to establish that they had been disarmed and that the ships would then be dispersed to a number of neutral ports. He would then lead the battleships personally into their designated port, arrange for the reduction of the crews to a care-and-maintenance level and then perhaps return to Germany. The general belief was that peace would be concluded by Christmas.

In his book Reuter said he decided that he would take the ships across if no other Admiral could be found, but initially he thought of coming home as soon as possible after the proceedings in the Firth of Forth. 'However, soon after the onward voyage to Scapa Flow, the feeling quickly came to dominate my mind that I should prepare a way out of internment worthy of the High Seas Fleet.' He anticipated a British 'betrayal' (the choice of Scapa rather than neutral harbours) but felt it had to be an accomplished fact before he could act on it. 'The betrayal would give us back our freedom of action; we could then do what we liked with our ships, we could also sink them.'

There is much talk of betrayal in public and private German records of the time, reflecting the entirely understandable bitterness in defeat of a proud but outnumbered nation which later was perverted by Hitler into the 'stab in the back' theory. This is not the place to go into the rights and wrongs of the Treaty of Versailles or the Armistice; to retain a sense of proportion one needs only to remember that the Germans had exhibited the same level of generosity to the Russians in making them sign the Treaty of Brest-Litovsk in 1917, which ended the war on the Eastern front.

Events continued to move at breakneck speed in Wilhelmshaven. Reuter took command of the ships at noon on 18 November and issued this written Order of the Day to the rebellious crews: 'I have with effect from today assumed command of the Transfer Formation; I know I am at one with the crews [in] that every man on this transfer voyage will so fulfil his duties that the Fatherland will attain an early peace.' On the same day, Reuter had meetings with Meurer and with the twenty-first committee of the Soldiers' Council which purported to run the fleet.

Meurer confided to Reuter that he was convinced the British would not allow the ships to go to neutral ports and that this had never been their intention, a suspicion fully justified by the facts from elsewhere. The Germans, including Reuter, always regarded this as a breach of the Armistice terms, in spirit if not in the letter.

They knew the wording of Article XXIII provided for internment in Allied ports failing neutral ones, but they felt very strongly that the Allies made no real effort to find any neutral harbours because they did not want to, and were thus guilty of a breach morally and in principle. The presentation of this kind of argument, plaintiveness cloaked in juridical syllogisms, is dear to the German soul and reams of paper were expended on it. In the end the decisive factor was that the Germans had lost a war, and in war more than anywhere else, might is right. Reuter recognised this in his book in an amusingly roundabout way, arguing that Germany had substituted Right for Might in its foreign policy, forswearing the latter in favour of the former, which is a neat if circuitous way of describing the defeat which left it no choice.

The Admiral had chosen a small staff of officers to assist him: junior Captain Ivan Oldekop (chief of staff), Commander Angermann, *Kapitänleutnant* von Freudenreich (interpreter), and two flag-lieutenants, Wehrmann and Tapolski. Backed by these men, notably Oldekop, for whom he could not find adequate words of praise, the Admiral set about the prolonged, delicate and dangerous task of reasserting a minimum of authority over the crews. His personal reputation as a first-class officer who had repeatedly proved himself when the war enabled and obliged him to put a lifetime's training into practice was probably his greatest asset in this formidable and barely successful task. Revolutionary acceptance of the necessity for internment helped too; there were even volunteers. Thus at the crucial first meeting with the twenty-first committee he won agreement on four extremely valuable points. Officers alone were to be responsible for seamanship; the Soldiers' Council was to co-operate in the internal running of the fleet; the officers were to have independence in running their own affairs; and the right of crews to refuse individual officers was withdrawn. This agreement was sometimes honoured more in the breach than the observance, but minimal officer-control, however precarious, was re-established at a stroke under the imminent hazard of internment. This did not prevent the committee chairman from observing grandly to Oldekop: 'Well, I have now taken over the command of the formation, and you are my technical adviser.' The filthy state of the ships, Reuter remarks, demonstrated the incompetence of the rebels. There was to be a lot of trouble with the crews, but the officers never lost on an important issue from this moment.

Another major event on the busy day of 18 November was the arrival in Wilhelmshaven of the battleship *Friedrich der Grosse*, lately flagship of the Fourth Battle Squadron, which was to serve as Reuter's flagship. In his report, Reuter remarked that this ship was the worst possible choice: 'I had put myself with my staff in a hornet's nest.' The great ship was in the grip of permanent pandemonium and it was impossible to understand how anything got done aboard. Wilhelmshaven, too, on the afternoon of 18 November as Reuter studied the orders for the transfer voyage, presented a picture of total confusion. Reuter was unable to establish whether all the ships had been properly and completely disarmed, whether they had the decreed amounts of coal, oil, food and water to last the four weeks until the Armistice was due to end or whether they were adequately manned. During the night of the 18th, supply trucks wandered aimlessly about the quays, not knowing where to offload, and radio and semaphore signals went from ship to ship in the frantic search for missing items.

Trusting to luck and hoping for the best, taking his courage in both hands, Reuter had a three-part order sent out by the High Seas Fleet command. The Transfer Formation was to be ready for sea at noon on 19 November, with steam up for twelve knots; the commanders of the Third and Fourth Battle Squadrons and the First and Second Reconnaissance Groups as well as the destroyer flotilla leaders were to meet him on the *Friedrich der Grosse* beforehand at 9 a.m.; and the revolutionary representatives, in effect the shop-steward of each ship, were to assemble one hour earlier on the flagship to elect a Soldiers' Council for the formation.

It was decided at the officers' meeting that Commodore Wilhelm Tägert would hoist his flag on the battleship *Seydlitz* and lead the formation, now officially classified as a unit detached from the High Seas Fleet, even though there would be precious little of that fleet left in Germany once the seventy-four warships left for internment. He would also be responsible for navigation and for ensuring that the fleet arrived at the exact rendezvous point (the same as was given for the *Königsberg*, fifty miles east of the Isle of May) dead on time. The disorder on the ships ruled out all but the simplest manoeuvres; the ships would therefore proceed to the overnight anchorage in the Firth of Forth designated by the British by turning aside ship by ship instead of squadron by squadron. Each captain would have sole responsibility for the seamanlike conduct of his ship: 'The Soldiers' Council has no joint say here,' Reuter ruled. The crews were to wear the Kaiser's blue uniform for the meeting with the Grand Fleet (an order which met with scant regard) and the decks at least were to be clear. Courtesy siren-signals were to be made only if the British did so first. Officers were to forgo special privileges in internment, such as the right to go ashore (which was never allowed) because these would only serve as a provocation in the circumstances. Some commanders reported that they were desperately short of officers, so much so that a number of captains were going to have to take watches in rotation with their subordinates.

The shop-stewards duly elected a Formation Soldiers' Council of three men chaired by Signaller Keller, who had never been to sea before, having spent the war as a shore-based signaller at the mouth of number III channel to the North Sea bases. Keller had smuggled himself aboard the *Friedrich der Grosse* with a forged letter on High Seas Fleet notepaper when the revolution broke out. He was the man who tried to put Oldekop in his place, showing his contempt for authority by the obligatory cigarette hanging from his mouth.

Unabashed, Oldekop told the new Council that there was to be no misunderstanding about the right of the captains to rule on all questions of seamanship. He also delivered a solemn warning about the import and likely consequences of flying the red flag on the ships on the open sea, as the rebels proposed to do. It meant, he said, that the ships could be regarded as pirate vessels and the enemy would have the right under international law to blast them out of the water on sight without warning. In his report Reuter duly warned: 'There will be more to say about the Soldiers' Council.'

CHAPTER SIX

The Last Voyage

REVOLUTION OR NO, the voyage to internment naturally aroused powerful emotions among the German sailors. The following passage is taken from a letter in the German Archives sent by an unidentified leading mechanic's mate on the battleship *Prinzregent Luitpold* to an acquaintance (he uses the polite *Sie* form of the second person pronoun rather than the familiar *Du*).

> What I have lived through in the last few days is hard to overcome for an old soldier who had always worn the Kaiser's uniform with honour. I lived through lovely days in the German fleet. I did not therefore shy away from taking part in the saddest voyage ever made by an undefeated fleet. That is the voyage of the German fleet to England for internment. The English had this stroke of luck only through the wretched, chaotic conditions in our Fatherland; the rats were not drawn from their holes for an honest battle, to their shame I may say.
>
> You will have read from the reports which ships took part in the voyage. All combat-tested ships from the Skagerrak battle. On the 21st we ran into the Firth of Forth under escort by English, French and American battleships. A good fifty large armoured ships, sixteen light cruisers, seventy torpedoboat-destroyers, countless torpedoboats, aircraft, airships and vessels of every kind accompanied us. It was more [of a] mockery for the great English fleet, for we could not after all

defend ourselves. Not a rifle, not a breech block and not a shell aboard!

A commission checked on our harmlessness. We could fly no Ensign. Three days we stayed in the Firth of Forth, then we went to Scapa Flow . . . You will also have read the reports of the English press. The drivel they put together about the achievements regarding our internment borders on idiocy. But that is of course the whim of the victor. There are no questions about the ways and means by which the victory was achieved. Perhaps a time is also coming when the cards will be shuffled differently!

The dishevelled capital ships, against all expectations, were ready for sea at noon on 19 November, with the exception of the battle-cruiser *Von der Tann*. She was late getting up steam because her signallers had been idle on the 18th and the sailing order from Reuter was not received until much later. Reuter therefore put back the departure to 1.30 p.m. Then it was discovered that the Second Torpedoboat Flotilla had not a single officer on any of the five ships which made up its Third Half-Flotilla. In the end it sailed without its officers. The Seventeenth Half-Flotilla was left behind to collect letters from the High Seas Fleet Command, and caught up with the rest of the fleet by 9 a.m. on the 20th.

With the capital ships in the lead, headed by the *Seydlitz*, the long line of vessels, once the pride of the Imperial Navy, passed the mouth of the Jade in what Reuter describes in his report as 'incomprehensibly fine weather'. They then steamed into the Heligoland Bight to the entry to Channel 400 through the minefields, made passable in the dark by specially positioned lightships. It was in this channel that the most dramatic incident of the crossing occurred. The destroyer *V30* (German destroyers had only numbers, like their submarines), carelessly steered, strayed off course and struck a mine. Two men died, the ship was abandoned and went to the bottom. At

6.54 on the morning of the 20th, Reuter radioed Beatty to report that only forty-nine destroyers would make the rendezvous (two days later, Beatty radioed Wilhelmshaven demanding a replacement). As they were passing Heligoland, says Reuter in his book, the officers were comforted by the fact that what one of them called 'the endless mourning-procession' now under way had ensured that the strategic island remained in German hands. 'Heligoland was worth this journey,' the Admiral wrote. The course of the fleet took the ships 'across the battlefield of 17 November 1917' where he had fought his last action so proficiently.

As soon as the ships moved into the open sea, the mutinous elements had second thoughts about continuing the voyage under the red flag. All ships thereupon hoisted the Kaiser's black and white ensign at their sterns, which put a little heart into the officers and many of the ratings too. A few red emblems were still to be seen elsewhere on several of the ships, but the danger of being attacked as pirates was eliminated. Reuter's report records the cowardice of the revolutionaries as they asked him for permission to change flags. Away from the coast, the weather turned cloudy and hazy with little wind. The *Cöln* (a light cruiser) reported by radio that her condensers were leaking, but she hoped to make the Firth of Forth. A sister-ship was ordered to stay close and tow her if necessary. As the Germans slowly sailed west, struggling to maintain a speed of eleven knots, they were bombarded with radio messages from the Grand Fleet asking for their course, speed and position. Reuter in his report rightly concluded that the British were following events in high tension, wondering whether the Germans had some last desperate gesture of defiance in mind.

The Germans were now in line astern in the order laid down by Beatty. In the lead were five battlecruisers. There should have been six, but one selected by the Allies, the *Mackensen*, was barely launched and not ready for sea. Then came nine battleships (there should have been ten, but the *König* stayed behind in Kiel with mechanical problems). Then came the light cruisers, seven instead of the stipulated eight because the *Dresden* could not be patched up in time – she had been badly damaged in action. Then came the destroyers in five groups of ten each (minus the lost *V30*), to make a total of seventy out of the required seventy-four listed in the second supplement to Article XXIII of the Armistice. The missing ships or replacements for them were sent direct to Scapa Flow later. The fleet, with gaps of three nautical miles between one group and the next, formed a line about nineteen miles long.

On the morning of Thursday, 21 November, all clocks and watches were put back one hour from German time to Greenwich Mean Time. There was still little wind, and visibility in the slight haze and the pre-dawn light just before 8 a.m. ranged between two and four miles. At that moment the beam of a searchlight was suddenly spotted from the battlecruisers. It came from a British destroyer and marked the first contact between the two great fleets. Beatty had told Meurer that 'a sufficient force will meet the German ships and escort them to the anchorage' in the Firth of Forth.

The British idea of a sufficient force staggered the Germans as ship after ship loomed out of the haze in two long lines, which soon began to pass down either side of the High Seas Fleet in the opposite direction. They finally put about to escort the German ships into the Firth as German officers and ratings with telescopes tried to count them. There were over 250 Allied warships in all, almost the whole of the British Grand Fleet, one squadron of American battleships as well as representative ships from other Allied navies. It was the largest assemblage of seapower in the history of the world. The mutineers noted, as the two fleets sailed due west at ten knots, that all the Allied ships to their north or starboard were flying red banners from their masts. If they thought this was the fondly imagined outbreak of fraternity on the part of the British sailors, they were wrong, as the ships on their port or southern side were flying blue ones. 'Operation Z2', for the reception of the German fleet, divided the Allied ships into 'Red Fleet' and 'Blue Fleet' for escort purposes, no more. Together they mustered thirty-four battleships, ten battlecruisers, two of the new aircraft carriers, about forty-six assorted cruisers and eight flotillas of twenty destroyers each. A sufficient force indeed. Airships and aircraft passed overhead. The guns of the British ships were aligned fore and aft as if for cruising, but their pacific appearance was deceptive. On Beatty's orders, the guns were free and not locked into position; their crews were on alert; the guns were empty, but the mechanical gunlayers were 'up and loaded ready for ramming home'. The gunnery control officers in their towers at the main-masts trained their fire 'directors' on the German ships, continuously calculating the closing range, changing deflection and adjusting the sights. They were ready to open fire in a matter of seconds and no chances were to be taken. With the escorts in position, Beatty, the flag he had flown at Jutland (or so Reuter was to claim) overhead, sailed up to the head of the triple line. As he passed each Allied ship, the ratings lined the sides and gave three cheers.

In an Order of the Day on the 21st, Beatty congratulated all ranks on

> the victory which has been gained over the sea power of our enemy. The greatness of this achievement is in no way lessened by the fact that the final episode did not take the form of a fleet action. Although deprived of this opportunity which we had so long and eagerly awaited and of striking the final blow for the freedom of the world, we may derive satisfaction from the singular tribute which the enemy has accorded to the Grand Fleet. Without joining us in any action he has given testimony to the prestige and efficiency of the fleet without parallel in history, and it is to be remembered that this testimony has been accorded to us by those who were in the best position to judge.

Fine words, redolent of anticlimax. They echoed a secret letter from Beatty to the Admiralty in the days while he was waiting for and organising the triumph, in which clear concern is expressed that not enough credit might be given where it was due (13 November). By all thinking persons, the part played by the British Royal Navy and the British Mercantile Navy in attaining this consummation is recognised, but it is likely that the full significance of the final victory at sea, obtained by the British, unmarked as it has been by any dramatic episode such as would appeal to popular imagination, may for that reason receive something less than its proper share of attention.

> It is due to our Empire, as well as to its first Service, that the character of the victory won at sea should be brought home to all by every means in our power. It is a victory that has no parallel in history. Crushing to our enemies, it has been finally achieved without a gun being fired or a life being lost.
>
> The enemy is required to hand over such a part of his fleet as will deprive him of the power of again contesting our sea supremacy, and this last act of his, which is the result of years of unsparing effort on the part of our officers and men and of ungrudging support on the part of our peoples, should in my opinion be arranged so as to afford an object-lesson, not only to our own countrymen and to those who live to see it but to all nations and to those who come after us.
>
> It will be for our good, and for the good of all that we as a race stand for, it will consolidate our position at the Peace Conference, serve as a recompense to seamen who have suffered at the enemy's hands, to seamen who have fought, to seamen who have waited, and be in some measure a tribute to those who have fallen, if the sea power of Germany is surrendered under the eyes of the fleet it dared not encounter, and in the harbours of the power that swept it from the sea . . .

> Advertisement of the deeds of the Navy is properly shunned by all naval officers, but the question to my mind touches great issues and its satisfactory handling will have widespread effects.
>
> The Navy can maintain its reputation for silence, but I would have the results of its long and devoted service displayed in a manner befitting an event so unexampled in our history so that all may see and remember.

Yet the arrival of an enemy fleet, disarmed by its own hand, in British waters was not lacking in drama of a clearly understandable sort. Nor as it turned out was this unprecedented moment the last act of the German fleet. Beatty's disappointment that the war at sea was not terminated by a decisive battle shines through this long, stilted plea of a letter. The scarcely concealed cheated feeling of the dashing Admiral is understandable, but the nation probably did in the end appreciate the good fortune which saved the lives of perhaps thousands of sailors and millions of pounds in damage. And Beatty certainly tried his hardest to make the best of the silent victory of the silent service.

The Admiralty also did its best to meet his wishes in a message to the navy of 23 November.

> The Board of Admiralty desire to express to the officers and men of the Royal Navy and the Royal Marines on the completion of their great work their congratulations on a triumph to which history knows no parallel. The surrender of the German fleet, accomplished without shock of battle, will remain for all time the example of the wonderful silence and sureness with which sea power attains its ends. The world recognises that this consummation is due to the steadfastness with which the Navy has maintained its pressure on the enemy through more than four years of war, a pressure

> exerted no less insistently during the long monotony of waiting than in the rare opportunities of attack.

The draft of this message bears a laconic comment by the American Admiral Benson: 'Good and dignified.' It was a great strategic victory, but there was no glory in it.

As they surveyed the colossal reception arranged for them, the Germans took it as a back-handed compliment to their own fleet's reputation. Reuter says in his book: 'The enemy could hardly believe that the victor of the Battle of the Skagerrak, this most feared German fleet, was really disarmed and what is more would not use this last opportunity treacherously to overwhelm the English fleet.' He thought that instead of cheering the British should have bowed their heads in shame in having the unbeaten fleet which had 'shattered England's historic mastery of the sea at Skagerrak' delivered into their hands.

Seaman Braunsberger on the *Kaiser*, tenth in the German line, wrote that an undefeated yet disarmed force like this 'is to be compared with an innocent man obliged to face the executioner'. As HMS *Cardiff*, a cruiser flying a captive balloon, passed the *Kaiser* on her way to take up station in front of the *Seydlitz* to lead the Germans into the Firth, Braunsberger, unaware of the internment programme, wondered to which neutral harbour they were meant to be heading. 'But we were badly disappointed, for . . . we reached the roads of the Firth of Forth and had to drop anchor.' Like several of his comrades, he recalls the guns of the escorts as having been trained on the German ships, but this trick of the memory probably arose from the fact that later in the day the ships were given a first examination by British officers to ensure they were completely disarmed. A more detailed check was made the next day. Nothing was left to chance.

Later on the 21st Beatty signalled the German ships in the Firth of Forth by radio: 'The German flag will be hauled down at 3.57 in the afternoon (sunset) and is not to be rehoisted without permission.' The German officers and the loyal members of the crews succumbed to a wave of emotion and bitterness which many of them were never to forget. Reuter made a strong verbal protest about this order as soon as he met the British, but to no avail. The ensign was not seen again until the scuttling. Vice-Admiral Friedrich Ruge, one of Germany's most distinguished sailors, who ended his career as head of the West German navy after the Second World War, was then a sub-lieutenant on one of the destroyers sent into internment. In his personal account, *Scapa Flow 1919*, Ruge regards Beatty's conduct when the German ships arrived as discourteous. He did not receive Reuter (indeed the two men never met) and Ruge wonders whether Beatty was expressing his abiding resentment over his 'defeat' as commander of the battlecruisers at Jutland at the hands of the German force of battlecruisers led by Hipper which was barely half as large. This is rather open to doubt, but Beatty sailed away to Rosyth soon after the Germans had entered the Firth of Forth and left all subsequent arrangements for the internment to his second-in-command, Admiral Madden.

The German ships kept station as best they could in the circumstances, but one or two had to sheer off out of the line to avoid collision, which made the British very critical and the German officers ashamed. The rogue ships had got up too much steam. As he surveyed the British ships around him, some 350 yards away on either side, Reuter (in his report) was impressed by the discipline and order he could see, in such heart-rending contrast with 'the thoroughly unkempt figures (who) lounged about on the decks of the equally completely unkempt German ships, who would not observe any order or any appeal to a last residue of a sense of honour below deck. Even the most portentous calls from the Soldiers' Councils naturally remained completely unobserved.' The *Cardiff* asked Reuter whether the Germans could make twelve knots. He agreed, but it proved impossible; the line spread out and lengthened, even his own flagship could not keep up, and the speed of the convoy as it entered the Firth slowed to ten knots. A request to try eleven had to be refused. The Germans were grateful for a change in the weather. The haze thickened and visibility deteriorated as they entered the Firth. Even in his official report (it is only to be expected in his book) Reuter allows himself an occasional expression of emotion: 'Thus Heaven conferred a certain mantle on our shame in the form of a light veil of mist, and the most tremendous tragedy ever enacted at sea was thereby softened to the view.' At the Isle of May, the port-side escort turned tail, falling in behind the starboard ships to form a single line which passed north of the island as the Germans passed south of it. Eventually, the German ships were ordered to drop anchor to a prearranged plan under British supervision off the island of Inchkeith in the Firth of Forth.

After the flag parade which saw the German Imperial ensign hauled down, Admiral Madden's chief of staff, Commodore Michael Hodges, two staff officers and an officer-interpreter boarded the *Friedrich der Grosse* in full dress uniform. Reuter received them on deck. Salutes were exchanged. He led them into his stateroom and invited them to take seats along one side of a long conference table while he and his staff took the other. An 'extremely cool' exchange of courtesies took place, whereupon Hodges handed over a thick packet of papers containing instructions for the German fleet. These went into the most minute detail about the conduct of the ships in internment, from the length of their anchor-chains to the level of pressure in the boilers. All radio transmission was banned, lighting at night was laid down and so were rules for the exchange of signals between the British and German fleets and within the German fleet. The German ships were forbidden to lower boats without permission. Although the most urgent aspects of these orders were signalled to the German ships by the *Friedrich der Grosse* as soon as possible, there was one incident that day when a German destroyer lowered a boat. Tension rose on the British destroyers alongside but the German explanation of ignorance was eventually accepted. As the British officers appeared on the deck of the flagship to return to their own, a crowd of German sailors 'in the most unkempt and filthy attire' (Reuter's report) gathered round the companionway, 'besieged it, lying on the railing-chains, squatting and kneeling, goggled at the Englishmen, shouted greetings at them and begged for cigarettes'. Reuter himself, using strong language, cleared a path for the British party, who showed their distaste at this self-abasement as they left. 'This picture cut deep into my soul and will never fade from my memory,' Reuter wrote.

He followed up his verbal protest against the order banning the German ensign (Hodges reported that it was made with 'great emotion') with a written one, arguing that there was no precedent for it; the ships should be in a neutral port and should in any case be treated as if they were. He appealed to the British sense of fair play and remarked that between chivalrous opponents it was not the custom for the victor to humiliate the vanquished. The British replied by saying that hostilities had only been suspended by the Armistice and a state of war continued to exist until peace was signed. No enemy ship could be allowed to fly its ensign while under surveillance in a British port. Reuter rejected this argument on the grounds that Article XXIII treated neutral and Allied ports as 'parallel'. But as usual the victor had his way. In a bitter little passage in his book, Reuter remarks: 'England knows by long experience that it ought to be kinder to people than to dogs, to cut off their tails bit by bit instead of all at once.' This reference to the old Continental practice of cropping the tails of certain breeds of dog is intended as an analogy for the way the British fudged the issue of where the German ships were to go for internment. Reuter says he first understood that the voyage to the Firth of Forth was to enable the British to check on the disarmed state of the ships; then he claims British officers expressed anxiety about the suitability of an anchorage in the Firth because it was exposed to the east wind; then Scapa Flow was presented as a more suitable anchorage in this respect; and then only after they had gone there did it become clear that this was where they would be interned rather than in neutral ports. Hodges claimed that he told Reuter unequivocally that internment would be in Scapa Flow. The German Admiral emphatically denies this, but the German record of their talks was lost, probably in the scuttling.

Reuter thought seriously, according to his report, of running up the German ensign on the 22nd as a gesture of defiance but soon had second thoughts. He had demanded strict adherence to British orders from the Soldiers' Council and he felt he could therefore hardly be the first to break them. He says he had already realised that he would not be able to avoid using British authority against the rebellious elements among the crews at some stage.

The extremely thorough check on disarming was carried out by the British on a one-to-one basis: a battleship checked each battleship, a destroyer each destroyer and so on, their guns cleared for action as the boarding parties examined everything for hidden weapons, including coal-bunkers and captains' cabins.

An account by Lieutenant Fritz von Twardowski, captain of the destroyer *G91*, survives in the German Archives. Despite the quaintness of the recorded English and the way the writer presents himself in a suspiciously heroic light, it manages to ring true (the quotations are in the English used in the original except where italicised, when they are translated from German by the author).

Twardowski says he had a wash and a shave. A British officer had already come aboard and did not like being kept waiting.

> I took my time, lit a cigarette and went on deck. I did not move towards him so he came to me. He had his hands in his jacket pockets, I had my hands in my trouser pockets.

The British officer asked Twardowski:

> 'You are the captain of this destroyer?'
> 'Yes, sir, and your position, please?'
> 'I am the captain of the *Speedy*.'
> 'All right, sir.'
> 'Where is your English officer?' [presumably 'interpreter' is meant here]
> 'Le voilà. What do you want, sir?'
> 'You will come with me round the boat!'
> '*I wouldn't dream of it unless you show a bit more politeness, old friend.*' [Twardowski here parenthesises that he frowned and raised his voice for the benefit of the German sailors watching.]

The British officer then said:

> 'All right, sir, would you be so kind as to answer me some question? [sic]'
> 'With pleasure, sir.'
> 'What's your name?'
> 'Von Twardowski.'
> 'How do you spell it?' [After the German officer had complied, he asked the British officer:] 'Do you take a cigarette?'
> 'I am so sorry, I am ordered to take nothing from you. (*He holds out a piece of paper.*) You are the captain of this destroyer?'
> 'Yes, sir.'
> 'What is its number?'
> '*G91*.'
> 'O, sir, this boat from Zeebrugge Flotilla? [sic]'
> 'Yes, have you ever been Niewport? [sic]'
> 'O, Yes, I know you quite well.'

'Net patrol?'
'O, awfully annoying and dangerous! Do you speak English?'
'I think so.'
'Would you be so kind as to lead me round the boat?'
'*Let's do that.*' [Twardowski concludes in German.]

The German skipper, like many another compatriot in the fleet, records that the British were most impressed by the quality of the ships and their armament. He got the impression, again like many another, that the British did not feel like victors, did not behave like them but rather resembled people who had had an amazing stroke of luck which they still could not quite believe. They asked whether Germany would start the war again on 17 December, when the Armistice was due to expire unless renewed. The British were 'terrified' of the revolutionary propaganda in case it proved contagious and strictly forbade fraternisation. But there was surreptitious contact between German and British sailors on the *G91*, for Twardowski later learned that the British regarded the captain of the *Speedy* as a fool and all the petty officers were resolved to leave the ship after the war because they had had enough of him. At the end of this ninety-minute preliminary search (a more thorough one was carried out the next day), as Twardowski records with understandable delight, the British officer's launch would not start for ten minutes, during which time he blushed furiously. *G91* was the only ship in the fleet not to haul down the Imperial ensign, Twardowski says; the confusion caused by the search had made the signallers miss the order. So a British petty officer climbed up and tore it down. The Germans laughed.

Twardowski detected little interest on the part of the local population in what was going on in the Firth of Forth. He noted one or two small boats with sightseers and added that King George V had been in the area earlier in the week and perhaps the locals had had enough excitement. Commander Hermann Cordes, leader of the torpedoboats and highly praised by Reuter, not only succumbed to the false rumour that the King was actually present on Beatty's flagship at the time of the Germans' arrival but fancifully adds that the sovereign had been led before 'the arrogant Sea Lord like the *Germani* before the throne of Nero'. In his own report on the events surrounding the internment, Cordes says the British suspected that the Germans would at a prearranged time and in response to an apparently inoffensive routine signal break out of the line, start shooting and ramming the escorts and then blow themselves up. This impression, shared by many German sources, is supported by the wariness of the overwhelming reception force Beatty assembled.

The official British effort to present their silent victory as an unparalleled military triumph rather than a strategic windfall was turned upside-down by the Germans. Cordes remarks that the Germans were 'betrayed' and the navy had no need to feel ashamed. They could look at the vast fleet which came to meet them and say to themselves how remarkable it was that they had held such an enemy at bay for over four years and frustrated him. There is surely some justice in this viewpoint. At this distance the argument is perhaps academic, but the residual pride of the German navy which it reflects helps to explain how Reuter was able to scuttle the ships despite his immense disciplinary problems. The general sense of betrayal – by the revolutionaries of Germany and by the British over internment – also helped. Ruge remarks that the enemy's failure honourably to carry out the few obligations imposed on him by the Armistice, as he sees it, paved the way for the scuttling.

With the chill ceremonial and the rivet-by-rivet double inspection at the Firth of Forth complete, the time had come for the German ships to weigh anchor and sail under escort to Scapa Flow. The second and final stage of the last voyage into internment was completed without major incident, but not without misgivings among the Germans. The order to sail on to Scapa Flow came to Twardowski as an unexpected further blow. The Germans knew of the main anchorage of their enemy only by repute, but what they knew was less than comforting. The thought of 'sailing another 250 miles, and then to a place where the wolves say good night to themselves' gave Twardowski no cheer at all.

Beatty had already radioed orders direct to the High Seas Fleet command at Wilhelmshaven on 20 November to send transports to Scapa Flow to collect those men of the German crews who would not be needed after internment began, but Reuter had no knowledge of this. As far as he was concerned, the British orders to him on 22 and 23 November to move the German ships in stages to Scapa Flow still did not reveal British intentions on internment. He claimed that his direct question to Hodges about the final destination on the 21st drew the answer that he (Hodges) did not know: 'he must have been lying'.

Two German destroyer flotillas left the Firth of Forth on 22 November; the rest went in two groups on the 23rd and 24th, each section escorted by an equal number of British destroyers led by command cruisers.

The larger ships started to move out on the 24th and their transfer was also carried out in three stages over three days to the 26th. Reuter himself went with the *Friedrich der Grosse* and the German Fourth Battle Squadron on the 25th, escorted by an equal number of British battleships, followed by three German light cruisers and their British equivalent escort two hours later. The *Cöln*, its condensers still playing up, limped in last. On 27 November, all the seventy ships which had gone to the Firth of Forth were in their appointed anchorage positions in Scapa Flow after passing through the Pentland Skerries and a triple boom-defence system. The British navigators placed aboard the German ships for the voyage were by all German accounts not over-endowed with competence and had to rely on German officers when manoeuvring into position.

The display of frustrated seapower was not quite complete: the battleship *König* after repairs arrived direct from Germany on 6 December. So did the still half-crippled light cruiser *Dresden*, hastily patched up and leaking badly, and the destroyer *V129* as replacement for the *V30* which sank in the minefield on the way across. The seventy-fourth and last ship, the magnificent battleship *Baden*, arrived on 9 January 1919 to stand in for the incomplete *Mackensen* which had been asked for by the Allies. Thus the German High Seas Fleet finally passed into the hands of the enemy, its guns useless, its great engines silent, its reduced crews in disorder, its ensigns stowed, its transmitters disabled, its future uncertain. '*Wehrlos, ehrlos*,' says Reuter's report: 'Disarmed, dishonoured.'

CHAPTER SEVEN

Internment

SCAPA FLOW IS, roughly speaking, a square bounded by land to the north, west and east and open to the south. Its eighty square miles of water served as the principal anchorage for the bulk of the British fleet in both world wars and the modern visitor can see evidence of this now abandoned role. The rusted hulks sunk as blockships in the First World War between the islands on the eastern side still exhibit their superstructures in many places and have been merged with the land by silting. These islands are now linked together by the Churchill Causeway, built by Italian prisoners in the Second World War, and carry a main road from Kirkwall, the Orkney 'capital' in the north-eastern corner of the Flow, to a point about seven miles north of John o' Groats. The southern coast of Mainland, the principal island, forms the northern shore of the Flow, which is bounded on the western side by the towering island of Hoy. The channel between Hoy and Mainland at the north-western corner of the Flow is overlooked by the harbour of the little stone town of Stromness. Guarding this channel from the inside is the rocky islet of Cava, east of Hoy and somewhat further south of Mainland. Due south of Cava and closer to the east coast of Hoy lies Fara, twice as large as Cava. Considering the Flow's recent importance as a naval base, it is surprising that other than the blockships there are remarkably few relics of this role. At Lyness on the east coast of Hoy are two long crumbling jetties, one of concrete, the other of wood, both too dangerous to walk upon. Ashore there are still some rusting oil-tanks, a few concrete remnants and the occasional scarcely legible notice at a drunken angle warning that you are trespassing on Ministry of Defence property. The place reeks of abandonment and disfigures the shore with its ugliness. Not far away along the one single-track road that straggles along the eastern shore of Hoy is the best-preserved remembrance of Scapa Flow's naval heyday, the beautifully tended naval cemetery in an enclosed field of lush grass. It is far from full, so that the double row of little grey granite tombstones bearing German names stands out, isolated near one end of the field from the rather larger parade of British graves at the other. Here are the remains of the eight Germans killed in the scuttling and four others who died from varying causes earlier. Fly over the Flow today and you are unlikely to see a single warship unless you happen to be there during the occasional naval exercise. Instead you will see a handful of supertankers moored near the island of Flotta, east of the southern end of Hoy in the southern mouth of the Flow, now a major North Sea oil terminal. On a sunny day, Scapa Flow is a starkly beautiful place; in bad weather, which is frequent, it is memorably desolate and a colour photograph comes out like a black and white print.

Although Scapa Flow itself offers a safe anchorage at all times and in all weathers, it is part of one of the most heavily populated maritime graveyards of all. This was true long before the Germans scuttled and has sadly remained so since. One or two ships of the Spanish Armada were lost in the area; so was Lord Kitchener, who went down with HMS *Hampshire* in June 1916 when she hit a mine west of Mainland after leaving the Flow. In July 1917, HMS *Vanguard*, a battleship, blew up while at anchor inside the Flow. A German submarine was sunk just inside the Flow in 1914, and in the early days of the Second World War another penetrated the defences and sank the battleship HMS *Royal Oak* at her moorings with enormous loss of life. Rather more recently, the Longhope lifeboat was lost with all hands on a rescue mission from its station at the southern extremity of Hoy. This history of death and destruction at sea reinforces the naturally sinister melancholy of Scapa Flow.

Such was the final destination of the pride of the German navy. The seventy-four ships which assembled there had cost about 880m Marks to build, some £44m in the money of those times. It is impossible to convert this sum into the inflated currency of today; in trying to evaluate the worth of the interned fleet, one can do worse than recall Churchill's remark in *The World Crisis* maintaining that each German capital ship was worth an entire infantry division of 10,000 men in terms of its demand on the national war effort. In terms of tonnage therefore, keeping the fleet of about 400,000 tons in all might have cost Germany twenty divisions, or a couple of armies, quite a diversion of effort even for what had been the world's greatest land power until 1918. We have already noted that the cost of *building* a capital ship has been put at the equivalent of raising an army corps of three divisions. The surrendered submarine fleet which had made a belated but much deadlier contribution to the German naval war effort cost 700m Marks.

The interned ships were anchored in the north-west corner of the Flow. The battlecruisers took the most westerly position between Hoy and Cava: the battleships and light cruisers were grouped between the battlecruisers and Cava, and in an arc north and north-east of the little island. The torpedoboat-destroyers were tied together in pairs and anchored in two parallel lines between Hoy and Fara, south of the capital ships. The British ships normally lay east of Fara and north of Flotta, which created communications problems. Cava lay directly between the German flagship *Friedrich der Grosse* and the British, and she therefore had to use the southernmost of the capital ships, the battlecruiser *Seydlitz* south-west of Cava, as a relay station for visual signal traffic with the British.

German recollections of the scene are distorted by the simple fact that neither officers nor men were ever allowed ashore in Orkney. The islands generally are remarkably fertile for their latitude and much of the landscape is surprisingly gentle, green rolling country. The great exception is the geological oddity of Hoy (the name derives from the Norse and means high island), which dominated the view from the ships. The almost sheer, black-brown scrub-covered hills of its eastern shore, loaf-shaped and often capped with cloud or mist, seem to rise straight out of the sea and in fact leave barely enough shore-space for the coastal road.

Reuter recalled the islands as 'mountainous and rocky'. The Germans experienced only one great storm during the seven months of internment but there were plenty of lesser ones. Scapa Flow 'makes a forbidding and grim impression on the visitor', Seaman Braunsberger of the *Kaiser*, anchored between Cava and Hoy, recalls: 'The bight was surrounded by treeless hills and was desolate and cheerless.' Twardowski on the *G91* expected to see large naval installations but spotted only a couple of floating docks, a few cranes and one or two tiny villages on his way to his anchorage. 'The life of the [Royal] Navy must have been thoroughly frightful,' he thought. He expressed his anxiety to a British officer. Surely the British did not expect a destroyer to stay in a place like this for months cut off from land? 'Why not? We are staying here [i.e. have been staying] for more than three months,' the lieutenant recorded the British reply in his usual shaky English.

General British response, especially in the Grand Fleet, to the horrified reaction of the Germans to their new surroundings was a mixture of grim satisfaction and mockery of their complaints. *The Times* on 14 January 1919 carried an article by an anonymous British naval officer which reflected these views: 'To the British naval officer, the fact that the Germans are being forced to live at Scapa Flow in their own ships is one of the most perfect examples of poetic justice in the whole war . . .'

The article however does go on to give a clue to the reasons for the German misgivings by recording the following exchange between two other British officers:

> 'If the Hun is squealing after four weeks at Scapa, I wonder what he would have done if he had had our four years of it?'
>
> 'You would howl yourself inside of four weeks if you had to stick it in a Hun ship.'

The Germans had every reason to anticipate acute discomfort in a prolonged period at anchor cut off from shore. The way of life in the German navy bore little resemblance to that in the Royal Navy, which was accustomed to inordinately long periods at sea (and did not forbear to complain). The German ships were not intended to be lived in and were not built for anything but very short periods of occupation (a fact which gave them certain structural advantages over their British counterparts – the bulkheads for example were much stronger because rather fewer doors were needed). The relationship between a German sailor and his ship was comparable with that of a soldier in an armoured regiment and his tank. The soldier is actually only inside the tank for purposes of battle, movement or exercises. He does not eat in it; he has to get out even to brew tea, and he certainly does not sleep in it. The German capital ships, with as little as one third the coal capacity of the British, were built for short forays relatively close to their home ports. In between times, they tied up at the quays and the crews messed and slept in barracks ashore. Thus only the most rudimentary facilities were provided aboard, including storage space, because the ships were not expected to go to sea for more than a few days at a time.

But for some of the internees, Scapa Flow did manage to offer occasional consolation. In his book, Reuter recalls the scenery as harsh and desolate more than once, but at the same time peculiarly attractive. The sunsets were spectacular, and every now and again the Northern Lights would appear in the night. He concludes his description of internment conditions with the words: 'There is still a God.'

Clarity on the fate of their ships officially dawned on the Germans on 10 December. The previous day, the High Seas Fleet

Command in Wilhelmshaven radioed to the C-in-C, Grand Fleet, this plaintive message: 'Where are the interned ships?' The reply, sent at 2.53 p.m. on the 10th, said: 'Vessels are interned at Scapa Flow.' The Germans protested. A message from Rear-Admiral Goethe, acting head of the High Seas Fleet Command, to the British C-in-C that same evening argued that the Armistice set neutral ports as the first choice for internment. German crews were being treated like prisoners against the clear sense of this provision. He made three requests: that the caretaker crews should not be treated as prisoners and should receive post and newspapers as soon as possible and without censorship; home leave should be granted free of charge on request from the command; and administrative communications should be delivered as soon as possible and uncensored. The British replied the following day: 'The ships interned are under surveillance which includes censorship of mails. The crews are not being treated as prisoners. Permission for officers and men to return to Germany will be granted as German transport facilities and circumstances generally permit.' There they were and there they would stay until the plodding Peace Conference decided otherwise. It was now up to the Germans to endure.

About 20,000 men had brought the ships to British waters and the British were impatient to reduce this high number, probably higher than it needed to be for the last voyage because of shore accommodation difficulties in Wilhelmshaven. But the German naval authorities had great difficulty in finding the necessary steamers to do the job. Finally the SS *Sierra Ventana* and the SS *Graf Waldersee* arrived with supplies on 3 December. The former was due to take twenty-five officers and 1,000 men and the latter 150 and 2,200. There were scenes of pandemonium in the mere six hours the two ships were allowed to stay in Scapa Flow, with men piling aboard on one side while supplies were carelessly unloaded, with enormous losses compounded by large-scale theft, from the other, and the British patrol-vessels added to the chaos by trying to hurry things along. The *Sierra Ventana*, hopelessly inadequate for the task, had to stay until 8 a.m. the following day as unloading was ineffectually completed. The decks of the *Friedrich der Grosse* alongside were piled high with disorganised mounds of supplies to be distributed round the fleet by the British. About 600 more men went home on the two ships than intended, making a total of about 4,000 who were taken to Wilhelmshaven. Two more ships, the SS *Pretoria* and the SS *Bürgermeister*, arrived on 6 December to collect 250 officers and 4,000 men and 250 officers and 1,500 men respectively. The *Pretoria* loaded crews who came from the Baltic Fleet based on Kiel while the smaller ship took North Sea Fleet crews back to Wilhelmshaven. This second stage of the reduction in crews was rather more orderly, as the two ships involved were better suited to the task. On 12 December the third and last stage began with the arrival of two further merchantmen, the SS *Batavia*, for 200 officers and 2,800 Baltic Fleet men, and the SS *Bremen* for the rest of the North Sea personnel – 500 officers and 1,500 men – to Wilhelmshaven. After more anarchic scenes, the two ships left Scapa Flow, escorted as usual by British warships into the open sea, on the 13th. One of the few points the Germans had won in their discussions and arguments with the British related to the size of the caretaker crews, which were more than twice as large as the British said they would leave aboard equivalent ships kept in port in a minimal state of readiness and repair. The numbers to stay on the interned ships were:

Battlecruiser:	200 (total 1,000)
Battleship:	175 (total 1,925)
Light cruiser:	80 (total 640)
Destroyer:	20 (total 1,000)
Overall total:	4,565 (excluding officers and warrant officers, who numbered about 250 in all).

In fact there is much confusion about the actual numbers left aboard in German and British official records and private memoirs alike, no doubt resulting from the confused circumstances prevailing at the time. The British made no attempt to count heads before or after the six steamers repatriated the bulk of the crews, and German sources give figures ranging from 4,700 to 5,900. When the British radioed Wilhelmshaven asking for further supplies (the Germans now had no working transmitters and all their messages were sent from the British flagship on duty to Wilhelmshaven; sometimes the British sent messages about the interned ships there without telling Reuter), they asked for quantities sufficient for 5,000 men. There was to be another major reduction just before the scuttling; in the intervening six months there was a steady trickle of repatriations caused by demobilisation, compassionate leave (usually a one-way trip), illness and punishment (usually rather welcome).

In his official report, Reuter states that the numbers interned were approximately 5,600 men and petty officers, 142 warrant officers and 175 officers. He does not explain the discrepancy between this total of 5,917 and the total of about 4,800 who should have been there to man the ships at the agreed levels (including the appropriate allowance for the four ships which came later to make up the total of seventy-four). Of these he says 1,435 men, forty warrant officers and sixty-seven officers came from the North Sea station and 4,223 men, 102 and 108 from the Baltic station, helpfully totalling 5,975. Yet in his own requests for supplies Reuter asks for rations and other items for 5,000 men in the early days of the internment, and one assumes the officers did not starve. An average of 100 per month went home during internment, and there were, according to evidence in the German Archives, about 4,400 men (excluding officers) on the ships until the last repatriation just before the scuttling, suggesting an original total of 5,250 officers and men! And so it goes on. It seems there were more than 5,000 at the start, anyway, and that a good 600 of them got home up to the week before the scuttle.

The feeding of the 5,000-plus was difficult enough amid all the indiscipline. The fact that the British insisted that all food should come from Germany throughout internment was a logistical headache, especially at the beginning when so much disorder prevailed in unloading from the freighters and steamers. Eventually a routine developed with fairly regular supply ships coming to Scapa Flow, unloading on to the flagship, from which supplies were distributed by the armed drifters and trawlers of the Royal Navy patrolling the interned fleet. In the end the crew probably gained from the arrangement that their own country should feed them; they may have eaten marginally better than their guards at times. The food was also a link with home, even though many people there did not eat as well as the sailors. But there were always shortages, especially in the early days. Ruge recalls living off iron rations which contained a high proportion of fat, all very well in battle when it was likely to be burned up quickly and for which it was intended, but inclined to make men in enforced idleness liverish.

On 10 December the British, without consulting Reuter, radioed Wilhelmshaven with a request for rations for 5,000 men for fifteen days or more, to be repeated every fifteen days. The message also referred to an acute shortage of soap, which persisted throughout internment, and other items such as writing materials.

The first supplies after those brought by the evacuation steamers came with the *Königsberg* on 21 December. The light cruiser, built for other purposes altogether, proved a highly incompetent supply ship. It took two days to unload from her to the flagship, by which time the goods were so hopelessly mixed up that it took the British patrol vessels another eight days to complete the distribution round the fleet. But the *Königsberg* also brought the next consignment on 9 January 1919 in company with the battlecruiser *Baden*, which arrived for internment the next day. Unloading supplies from the two warships proved even more complicated than on the previous occasion. From January onwards, the job was taken over by the SS

Dollart, a purpose-built naval supply ship, and the SS *Reiher*, a cargo steamer, and conditions improved enormously.

Reuter concerned himself with the details of housekeeping for the fleet, now referred to as the Internment Formation rather than the Transfer Formation, from the very beginning. He had an enormous shopping list drawn up which the *Königsberg* eventually took back to Wilhelmshaven on 23 December. The requests included fresh meat, vegetables and potatoes, and also long-lasting foods. For ten days he wanted 10,000 loaves, 250 kilos yeast and 2,500 litres rum (one thing the interned ships were usually not short of was liquor). The Admiral also urgently requested toothbrushes and toothpaste, matches, lighters, flints, bootlaces, dubbin, thread, seaboots, canvas shoes, spare soles and of course soap. He also asked for 420,000 Marks in cash to enable the men to buy drink and small personal items in their messes. He made frequent requests for tobacco, which were met in fitful fashion (one order was for one million cigarettes). It became possible however to supply a monthly ration of 300 cigarettes or seventy-five cigars per man. Yet Reuter's last message to Wilhelmshaven before the scuttling was a complaint about the shortage of cigarettes and a request for urgent supplies, on 17 June 1919. A thriving barter industry rapidly developed, not only among the men but also between ships. Signallers on the mighty warships exchanged messages via the flagship which might have come from the columns of *Exchange and Mart*. The battlecruiser *Von der Tann* urgently needed a tool for a repair job: could any ship supply it on loan?; the battleship *Hindenburg* advertised that it had 150 rolls of lavatory paper to spare, while the *Kaiser* mentioned four capstan-bars and the *König* generously offered 195,000 surplus cigarettes. There were repeated requests for rat-poison; the neglected condition of the ships had encouraged the rodents to run wild.

Odd little obsessions arose from the shortages. A paymaster wrote to a comrade in Berlin from one of the capital ships: 'Life here is dreadful . . . I recently received a white cabbage from

Germany, and I tend it like a flower in my cabin.' Illicit barter also and inevitably developed with British sailors on the patrol-vessels, despite the severe penalties for any form of fraternisation threatened by the Royal Navy. There were even cases of unauthorised contact between the two sides at night, when British ratings 'borrowed' small boats and rowed out to the German ships, which were only a few hundred yards offshore. A British officer accepted a bottle of schnapps for some old newspapers; a German officer offered his Iron Cross for two bars of chocolate and sailors exchanged items of uniform, to the disgust of the officers, for soap and other articles. The Germans found an inexhaustible appetite for souvenirs among their guards.

The general supply position improved from 10 January 1919, when the High Seas Fleet Command was wound up and a supply office for the interned fleet was established in Wilhelmshaven. Many of the ensuing shortages were caused by the 'disappearance' of supplies between loading in Germany and delivery in Scapa Flow. Meanwhile the Soldiers' Council on the interned ships made objections to the alleged special treatment afforded to Reuter's mess for himself and his staff. In fact a medical officer who examined the Admiral diagnosed chronic intestinal catarrh arising from the unbalanced nature of his diet, which was low in vegetables and brown bread while containing plenty of meat and too much white bread and cakes.

Given the circumstances, the medical officers generally found that the overall standard of health remained surprisingly good in spite of the supply difficulties, the low level of activity aboard the ships and the complete ban on going ashore. On deck the air was unmistakably healthy. There was only one important and chronic health problem: the condition of the men's teeth. There was not a single dentist in the fleet, nor were the British prepared to provide dental facilities ashore. No dentist volunteered to go into internment before the ships left Germany. The monotonous diet was largely responsible; Ruge reports occasional cases of scurvy, once the curse of all navies. A special problem was the number of men with false teeth which broke and could not be repaired except in Germany. The staff doctors made a list of such cases but all the medical records were lost in the sinkings. In the end men with serious dental problems were sent home on leave with the returning supply ships. Reuter's repeated requests to Wilhelmshaven for a dental surgeon were, however, finally met. On 20 June 1919, a dentist called Grote volunteered and signed a contract. He was to set out for the interned fleet on 23 June – just two days after the ships went to the bottom. Thus sailors with rotten teeth or broken dentures became a significant proportion of those sent home on leave.

Reuter includes in his report an account of a curious accidental encounter by two internees, not previously acquainted, while they were on leave. A lieutenant-commander in civilian clothes met a stoker from the *Seydlitz* in uniform, identifiable as such from his cap-band. On looking more closely, the officer discerned a curious medal-ribbon on the man's chest. It was the black, white and red ribbon of the Iron Cross, but the broad black stripe down the middle of the ribbon had been crudely painted red. These ribbons often bore a clasp in the centre to indicate how the medal had been earned (for submarine or aircraft service or for a wound). The officer saw an obviously home-made clasp in the form of a crudely executed armoured cruiser at anchor. 'What's that?' he asked.

'That is the Order of Transfer into Internment,' the stoker proudly replied.

Many who got leave naturally set off with no intention of returning. This applied particularly, of course, to those being sent home for court-martial or punishment, who found it easy to evade both by vanishing into the postwar disorder at the ports and inland. But those without such over-riding reasons for deserting often changed their minds and returned on seeing the sorry state of the Fatherland, the political unrest and the social and economic disarray. Some concluded that by staying they would only become a burden to their families, who at least received a regular allowance from the navy as long as they were away. When older or long-serving men left for demobilisation leave, at Reuter's request they were not replaced because he feared that more agitators would infiltrate the ranks. He had good reason. Agitation was rife among the volunteer crews, scraped together with much difficulty, on the four replacement ships which had joined the interned fleet later.

Reuter himself went home on what turned out to be protracted leave aboard the SS *Bremen* on 13 December 1918. He too considered the idea of not coming back, not out of any intention to desert but because he was clear in his own mind that he had committed himself only to delivering the ships into internment and seeing them to their anchorage. The impending journey home was therefore 'synonymous with my retirement from the formation', he wrote in his book. But second thoughts arose: 'I had meanwhile grown used to my work and had discovered a circle of willing helpers.' On reflection he decided that the interned fleet did require the presence of an Admiral and that his duty must be to preserve what he could of the navy and to ensure if he could that the ships remained capable of going home if that were ever allowed.

Leaving Commodore Dominik, the recently promoted captain of the battleship *Bayern*, to deputise for him, Reuter boarded the *Bremen*. It was a depressing voyage. The crowded conditions on the ship with its complement of repatriated sailors were made even less comfortable by the tension between officers and ratings, who would allow their unwanted superiors none of the special privileges hitherto afforded to their rank. Then the *Bremen* ran aground off Wangeroog, one of the chain of East Frisian islands strung out along the German North Sea coast. The ship was stuck fast for half a day until the next high tide floated her off undamaged. Reuter does not have much to say, either in his book or his report, about what he did while he was on leave between his arrival in Wilhelmshaven on 14 December and his departure from the same port on 25 January, beyond remarking that his return was considerably delayed by conferences with his colleagues and superiors and the necessity of improving supplies to the interned ships. His family were in Wilhelmshaven at the time, as they had been for most of the war, and no doubt he spent as much time as possible with them, though he makes no mention of the fact in his writings that survive and deal with the internment. In the light of subsequent events the veil he draws over this period, whether accidental or not, is intriguing. After the scuttling the British tried everything they could think of to establish that Reuter had done it on direct and specific orders from Germany, as we shall see. It is quite simply inconceivable that in the staff meetings to which he so briefly refers there was no discussion about the ultimate fate of the bulk of Germany's surface fleet, to be resolved by the Peace Treaty then being negotiated, even though there was no clue to Allied intentions at that time. The subject must have been uppermost in every serving German naval officer's mind. It is a subject to which we shall return.

The British laid down strict rules for the conduct of the interned ships in a document entitled 'Instructions for German Ships' (IGS). Not only was there a permanent ban on flying the German ensign and on radio transmission (supplemented in February by a ban on radio receivers to cut off the mutinous elements from 'propaganda' from home), but the ships were to be maintained in such a condition that they could get up steam to manoeuvre and cope with bad weather (and to power generators for heating and lighting). Each ship was to show three lights at the bow and two at the stern after dark (strictly enforced: boarding parties were made ready when a ship failed to comply). The sole channel of communication between the German formation and the British authorities was to be from Reuter to the responsible British Admiral (there were five during the internment period). Outgoing post was to be censored (later incoming mail was also. Censorship was originally carried out at Scapa; but later it was done in London, causing an average additional delay of ten days). German newspapers were also censored. Boat traffic between the German ships was forbidden and anyone launching a boat without permission risked a British order to open fire.

The British provided two drifters for traffic between the German flagship and other interned ships and a third for exchanges between the German flagship and its British counterpart. No personnel were to be moved from one ship to another without permission: the only Germans who enjoyed the run of their own fleet were the two chaplains, Pastor Ronneberger (Evangelical Lutheran) and Father Esterkand (Roman Catholic), who took maximum advantage of their freedom and proved surprisingly popular, even among the rebels. Each ship was to report once a week through the German flagship on the state of the crew, coal, oil, water, food and other supplies. No ship was to move without permission and the ordained plan of anchorage was to be strictly adhered to. Ships without their own medical officer, such as the destroyers, were to be covered by surgeons from those which had. The IGS regulations ran to a good 10,000 words, not counting the occasional amendments issued by the British, and copies, badly duplicated in purple ink on poor-quality paper, are to be found in both the German and the British records.

No less detailed were the orders from the British command to Royal Navy personnel at Scapa, under the heading 'German Surface Ships' (GSS). These provided for the enforcement of the regulations imposed on the Germans. They also anticipated an attempt to scuttle from the very beginning of internment: guard-vessels were to report at once 'if a German vessel appears to be settling down or to be sinking at her moorings'. The surveillance at close quarters was to be done by a total of eight drifters or comparable vessels, three to be on duty at a time.

The drifters were to be relieved in rotation every twenty-four hours at 10 a.m. The crews were civilian under a merchant navy or fishing fleet skipper, but each boat had an armed party of sailors and marines under a naval officer aboard. 'Fire will not be opened on any boat or vessel without the authority of the officer in charge.' If a German ship slipped its moorings and tried to get away, the Royal Navy warships on duty in Scapa Flow were to prevent her escape. If the German ship reached the open seas she was to be ordered to return. If the order was not promptly obeyed, a warning shot across the bows would follow and if that did not work, the runaway ship was to be sunk by torpedoes and gunfire. Boarding parties were also kept ready on the supervisory British ships to board any ship to protect German officers and restore discipline if necessary (this regulation was hardly ever invoked, although there were several occasions when it might have been). Each party consisted of three officers and forty-five seamen, stokers and marines equipped with two Lewis guns, cutlasses, rifles with bayonets and ten rounds of ammunition, and revolvers. Each capital ship on guard duty had such a party ready; each light cruiser had a smaller one of two officers and thirty men similarly armed. One ship in each category was to stand guard in rotation for twenty-four hours starting at 8 a.m. In the event of boarding, a destroyer was to close the German ship and to be ready to fire its guns and torpedoes in the event of resistance. One light cruiser – or failing such, a battleship – was to have steam up for eight knots at one hour's notice and sixteen at two and a half hours as 'emergency ship'. Officers were not to stray more than one hour's travelling time away from the emergency ship.

Not only was fraternisation strictly forbidden to the British at all levels, but all ranks were reminded of the 'strictly formal character' of relations to be maintained between the two navies. 'It is to be impressed on officers and men that a state of war exists during an armistice . . . In dealing with the late enemy, while courtesy is obligatory, the method with which they have waged war must not be forgotten.' And again: 'It is anticipated that the attitude of the officers and men of the German ships which are to be surrendered will be friendly . . . Our attitude is to be courteous but not even distantly friendly.' Handshakes were barred and no British naval personnel were to accept food or eat aboard a German ship.

The IGS and GSS orders were drawn up by Admiral Madden, who had been delegated by Beatty to organise the entire internment operation. Shortly after the Armistice the Royal Navy was re-organised and the Grand Fleet was split up, never to form again. Madden became Commander-in-Chief, Atlantic Fleet, but retained overall responsibility for the internment throughout its duration. He took direct personal charge for the first few weeks, and then delegated responsibility to the Vice-Admiral commanding one of his squadrons of capital ships on a rotating basis. Each squadron did guard duty for about one month and by the end of internment five Admirals had served as 'Senior Naval Officer Afloat, Scapa'; one did two tours of duty.

Reuter too drew up a long series of orders for the organisation and administration of the interned ships, all of them counter-signed by the Soldiers' Council of the Internment Formation. The difficulties of working with the Council were eased somewhat by a decree of the newly established Reich Government of 19 January 1919 which went a considerable way towards reducing the authority of such councils and restoring some authority to officers. Sensibly however he felt it would be unwise to seek special privileges for officers. This self-denying ordinance probably cost the officers the chance to go ashore in Orkney, which seemed to be a possibility when Reuter had talks with Madden.

But many of the British restrictions were welcome to the embattled Admiral, particularly those designed to prevent contact between the different crews, even though the denial of shore leave provoked in him 'a sense of humiliation'. He was concerned for the health of his command and never tired of requesting permission, though the British never relented. At first Reuter was allowed to use his own pinnace during daylight to inspect his ships, but he soon gave up this privilege because the Soldiers' Council infiltrated its crew and used the inspection trips as a useful means of communication with agitators on other ships. The Admiral felt obliged to make his journeys in one of the British drifters detailed daily for the purpose. He maintained contact with his captains by having them brought to his flagship by the same means after giving due notice to the British. He ordered all ships to maintain themselves in such a way that they could get up steam for ten knots, to be able to go on to neutral ports or even go home, and also to be able to manoeuvre in storms or in case an anchor-chain or a mooring-cable broke.

He also drew up a skeleton command structure for the ships, reducing the number of officers, and therefore the chance of friction with recalcitrant crews. The capital ships were to be under the orders of a commander normally, with two seaman officers, two assistant engineers, one surgeon and one administrative officer or paymaster. The light cruisers were normally captained by a commander, a *Kapitänleutnant* or a senior lieutenant; with one executive officer, one assistant engineer, a medical officer and an administrative officer. He left the organisation of the torpedoboats, several of which had arrived with no officers aboard, to their leader, Commander Cordes. He based his organisation on the fact that his ships were anchored in pairs tied together.

The Soldiers' Council supported Reuter in his order that all British regulations should be strictly observed and that all contacts with them should be 'correct', a rule which was not often observed, except by the officers, who were encouraged to maintain a politely hostile attitude rather than a 'precariously friendly' one. At the same time, Reuter set out to win the confidence of Madden in order to minimise the risk of British interference in the internal running of the fleet, which could endanger his guiding principle that nothing must be allowed to happen that called into question German sovereignty over the ships in internment. He was at pains when he met the British commander to explain that the officers opposed the revolutionaries to a man, and British suspicions about their motives and sympathies were without foundation. Madden was convinced and the two men found the time to develop a degree of mutual respect. Reuter notes that the British Admiral personally conducted him to the companionway when he left the British flagship to rejoin his own on 27 November. 'The formality observed towards me was of a cold nature but not without a certain courtesy,' he wrote. He issued an order banning propaganda, again with the motive of not provoking the British. He noted with satisfaction that although the German war flag could not be flown, the British allowed the command flags and pennants to fly on the German ships, a satisfying acknowledgment in his eyes of German ownership and sovereignty aboard the interned fleet. The British proved punctiliously correct in this respect and their surveillance, however strict, remained external and they were always aware of the distinction between internment and surrender, even though they knew that not placing British guards on the German ships would make it very difficult to prevent them being scuttled. Reuter had no complaint about the distant relations he had with the squadron Admirals delegated by Madden to Scapa Flow (Admirals Pakenham, Oliver, Leveson – two tours – Keyes and finally Fremantle with whom he had a stiff exchange about the scuttling).

Inevitably, some kind of routine developed, differing widely from ship to ship and between one class of ship and another, as the crews sought their own ways of coming to terms with the sheer boredom of internment afloat.

Conditions were physically least uncomfortable on the big ships which offered the most space and coped best at anchor in rough weather; against this, the atmosphere on many of them of discontent, agitation, indiscipline and filth caused by neglect did not serve to make them desirable residences. The opposite was true on the destroyers, where acute physical discomfort was counter-balanced by relatively high morale. The intimacy forged in battle on small ships such as torpedoboats (and notably lacking in the capital ships), the essential trust between officers and men based on a mutual desire for survival and the need for teamwork in danger, and the sound leadership of Commander Cordes soon brought relative stability. In his report, Cordes said that the cluttered decks of the torpedoboats permitted a man to walk a maximum of forty paces in one direction along a strip of deck five feet wide. He could recall the weather allowing use of this facility just twice in March 1919. The little ships began to roll in sympathy with a wave before it actually reached them. The crews did the necessary work efficiently enough if not exactly with gusto. Soldiers' Council representatives on the destroyers were of a distinctly more moderate stamp than elsewhere and adhered to the limited role given them by the Reich Government decree of January 1919. Sometimes, says Cordes, they proved positively useful; in general, the officers got their orders carried out. A secret ballot of the 1,000 destroyer men produced 862 votes in favour of the new government and Reuter and only forty-four against. The fleet as a whole declared its loyalty to the new regime. Of their own accord the crews collectively decided to offer their services en bloc to the new government when they got home and to stick together as a unit for festivals, old comrades' gatherings and the like after it was all over. Soon after internment began and before the main supply problems were resolved, the destroyers ran out of oil for their turbines and their crews were temporarily evacuated to the big ships. The men were appalled by what they saw when they got aboard; the destroyer man's traditional contempt for big ships was doubled and redoubled by the appalling, self-inflicted conditions.

Although the British insisted that all provisions and general supplies from buttons to spare engine parts should be delivered direct from Germany, they recognised the insuperable difficulties attached to requiring the Germans to find their own coal, oil and water. The German ships, unlike the British, were not equipped with desalination plants for the distillation of drinking water from seawater, an expensive item which justified itself only for ships intended to be at sea for a long time. The sheer bulk of the coal needed just to keep the larger ships ticking over and of the oil for those with diesel engines would have required an endless convoy between Wilhelmshaven and Scapa Flow, logistically out of the question in postwar circumstances. So the British agreed to supply the interned ships with these commodities, and water, presenting the bill to the German Government. The delivery of coal to the ships at the offshore anchorages created problems of its own. The Germans were accustomed to bunkering at coaling points in port. It was far from easy to transfer sacks of coal from a small supply ship to a large warship in a choppy sea in the limited hours allowed by the British. On 8 February 1919 the Soldiers' Council protested to the Reich Naval Office, which now ran the German navy, and to the German Armistice Commission about what they saw as the British coal swindle. They claimed that if, say, 500 tons of coal were to be delivered to a given ship on a given day and only 200 tons were in fact supplied because of defective loading gear or bad weather, the British still charged for 500 tons at 80 Marks a ton. Even though the crews cut their lunch-break to fifteen minutes, the British deliberately slowed down the loading process. A report from the Interned Formation Command confirmed some of this but threw a different light on the problem on 30 January. The German crews worked badly in loading the coal, it said, so that supply ships had to lie alongside much longer than necessary at greatly increased cost. A loading 'norm' was set by the British so that the specified amount of coal was deemed to have been delivered in the time set even if it were not, and there was to be no supplementary delivery to make up the shortfall. But only the true amount supplied was actually charged for. Things improved with practice, but every now and again the crews had to sit in the dark wrapped in their greatcoats and blankets as the engineers sought to eke out the coal stocks to the next delivery. To conserve oil, the destroyers were subjected to 'lights out' at 11.30 p.m. when their diesels were turned off. The same report said that the British system of distributing general supplies round the fleet was working well and the Royal Navy had managed to find a water-tender, something it had never needed to do before. This was the *Flying Kestrel* from Liverpool, which took eight days to make a complete round of the interned ships, and which was on hand at the very moment the ships were scuttled.

In terms of international law, the crews aboard the ships were internees, not prisoners. Quite understandably, they could not see the difference and their plight produced psychological effects indistinguishable from those observed among prisoners of war on land or convicted criminals in civil jails. Several who lived through internment referred variously to 'barbed-wire sickness' and 'the Scapa Flow syndrome'. The form it took varied with the personality of the individual; the symptoms included melancholia or depression, obsessive behaviour, aggression, irrationality and apathy. The common denominator was boredom. Seaman Braunsberger recalls:

> The further spring advanced, the more monotonous it became for us. We received little post, no newspapers and we could not receive news over the radio either. The reluctance to work got worse. Engine maintenance was no longer properly carried out, the water for the boilers turned salty. The power was often cut so that we had to sit in the dark in the evenings. Lunch often had to be cooked on the coal-burner and the diet got worse. Always the same comrades round you, always the same ships round you, always the same view of land etc. On top of that the agitation of the Workers' and Soldiers' Council which sowed only hate and strife. Even though we still believed in a happy ending, all this turned us into psychopaths.

As mentioned earlier, the atmosphere varied widely from ship to ship. The main difference lay between the smaller ships and the capital ships. Reuter and other sources agree that the psychological climate was worst of all on the five battlecruisers, the largest ships with the largest crews. Things were very nearly as bad on several of the battleships, though the Admiral remarks that Braunsberger's *Kaiser* remained reasonably orderly to the end, the best of the big ships. The Soldiers' Councils established dictatorial rule on most of the large vessels, and the situation for the officers aboard the battle-cruisers *Von der Tann* and *Derfflinger* was 'hair-raising'. On balance, the best place to be was probably on one of the light cruisers which did not bucket up and down like the destroyers and were generally not infested with malcontents and agitators like the capital ships. Reuter reports that the light cruiser *Bremse* was exemplary, run just as in the old days before the unrest crippled the fleet. Officers were piped aboard and the men put on their best uniform for the Kaiser's birthday. Reuter thought of restraining the zeal of the captain, Lieutenant Schacke, even though he obviously enjoyed the support of most of his men, for fear of provoking the crews of other ships. A plot by extremists aboard the *Bremse* itself to kill the captain was uncovered. The *Emden*, a light cruiser where the captain and the Soldiers' Council co-operated fully in running the ship, was also exemplary. Discipline was at its worst on the ships of the Third Battle Squadron from the Baltic Station at Kiel, where revolutionary influence had been strongest and most uncontrollable.

Of his own flagship, the *Friedrich der Grosse*, Reuter says in his report: 'The ship was a madhouse.' As far as he was concerned, the battleship *Kaiserin* was dominated by 'evil elements' and the *Grosser Kurfurst* took the prize as the most slovenly ship of them all. Reuter eventually got some peace only when he moved his flag from the *Friedrich der Grosse* to the *Emden*.

There was, after all, little enough to do. Originally the men were supposed to work five hours a day with a break in the middle, on maintenance and cleaning, stowing supplies and the like. Quite often the work amounted to only two hours a day, and sometimes there was nothing at all for men to do. There were books, one could write letters, sew, draw and make things out of metal, play chess, get drunk when a new allocation of liquor arrived or stare vacantly at the same old view. Music and singing, sometimes organised, sometimes spontaneous, were popular and so was dancing. There were plenty of musical instruments aboard, and the British were amazed to see entire ship's bands appear on the decks of some of the ships on 9 February to play in honour of Reuter's fiftieth birthday. Gambling was rife, especially on the big ships. Men played *Skat* and other card games for high stakes and helped to recycle the generous quantities of money supplied to the ships which could not be spent anywhere else and which otherwise could be used only to buy small personal items or liquor (beer cost 50 Pfennigs for half a litre), all sold at low fixed prices.

The Supreme Soldiers' Council of the Internment Formation was rarely idle. On 19 January it drew up a series of demands which were taken by a spokesman repatriated on a returning supply ship and presented to the Reich Naval Office. One demand, the official recognition of the Supreme Council based on the flagship, was granted on 24 January and meant that the Council outlasted all the others in Germany itself, where they were abolished in March and replaced by elected spokesmen or shop-stewards. Also recognised were Soldiers' Councils – three men for each capital ship and destroyer flotilla plus a spokesman for each destroyer. The other demands on the list, granted virtually in full, included: dispersal of the interned crews to be subject to the consent of the Supreme Council (consultation was promised); establishment of a special supply office for the interned ships (already done); priority for professional sailors of the interned crews in the postwar navy (Berlin said it would do its best); 5 Marks a day internment allowance for all ranks including officers (they were awarded two Marks a day

Admiral Ludwig von Reuter, commander of the interned German ships.

Kaiser Wilhelm II, in the uniform of a Grand Admiral.

Admiral Tirpitz, creator of the High Seas Fleet.

Admiral Scheer, commander of the High Seas Fleet at Jutland.

Admiral Beatty, commander of the British battlecruisers at Jutland and later of the Grand Fleet.

Admiral Jellicoe, commander of the British Grand Fleet at Lutland.

Admiral Hipper, commander of the German battlecruisers at Jutland, who later succeeded Scheer.

Top. HMS *Queen Elizabeth* leading German battleships into internment.

Above. SMS *Friedrich der Grosse* (foreground) and other German battleships at Scapa Flow.

Top. SMS *Seydlitz*, leader of the German battlecruisers.

Above. German torpedoboat-destroyers at anchor in Scapa Flow.

The interned fleet seen from the north, with light cruisers at left and capital ships in the centre and at right.

A painting by Cecil King of a British drifter (left) patrolling the interned fleet.

Top. SMS *Frankfurt*, a light cruiser, awash after scuttling.

Above. SMS *Derfflinger*, a battlecruiser, turning turtle.

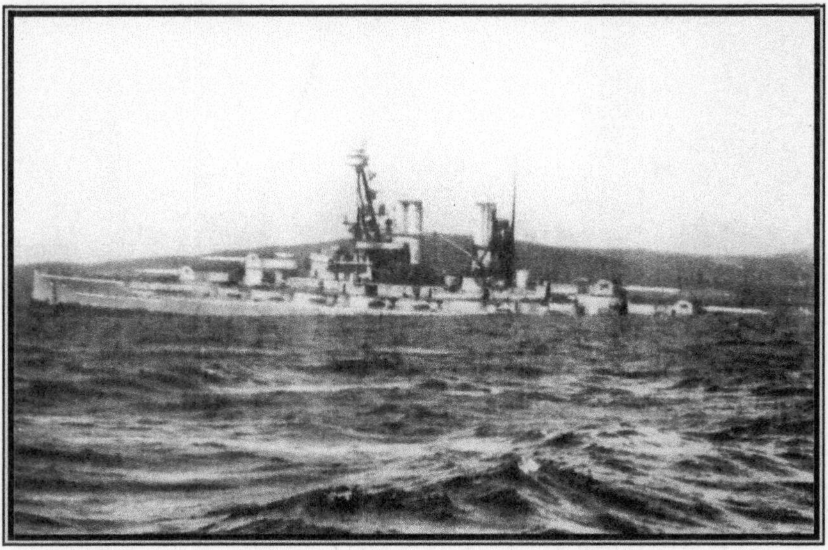

Top. SMS *Baden*, a battleship, awash (right) with *Frankfurt* in the background.

Above. SMS *Bayern*, a battleship, sinking by the stern at 2pm on 21 June 1919.

Top. The *Hindenburg* on the bottom after sinking on an even keel.

Above. The *Derfflinger* after coming to rest on her upperworks (she was often mistaken for an island until her salvage).

German sailors in a lifeboat surrendering to a British ship, with SMS *Nürnberg* settling in the background.

Admiral Fremantle denounces Admiral von Reuter on the deck of HMS *Revenge*. This photograph was taken by Bernard Gribble.

Top. The *Prinzregent Luitpold*, with air-locks attached, alongside a floating dock.

Above. The *Moltke*, after delivery to Rosyth for breaking up.

plus a special single payment of 225 Marks which included a premium of 100 Marks for taking the ships into internment); a gratuity and 'demob suit' on completion of service (granted); four weeks' leave on full pay and allowances on return (agreed provided internment lasted at least four months); an increase of 50 Pfennigs a day in subsistence allowance (agreed). The monthly payroll for the ships was over 500,000 Marks plus about 35,000 Marks paid to relatives in Germany.

When Reuter embarked on his long leave in the middle of December, he hoped that calm would descend on his interned ships as the crews got used to their new role and the initial difficulties were overcome. When he returned with a supply ship on 25 January, he found an atmosphere of crisis, with Commodore Dominik embroiled in a struggle for power with the Soldiers' Councils. The confrontation endured for most of the first quarter of 1919 and only ended when Reuter changed his flagship.

The roots of the protracted tug of war lay in the irreconcilability of the officers and the Soldiers' Councils. It was agreed that they should share command, but it was inevitable that one party or the other would emerge supreme, as any command thus divided and founded on mutual mistrust could not function. The spark was the news of the deaths by shooting in a demonstration in Berlin of Karl Liebknecht and Rosa Luxemburg, the two idols of the revolutionary ultra-left, which reached the interned crews on 16 January. The Soldiers' Councils organised a drunken wake, breaking out enormous quantities of liquor. The drinking went on until four the following morning and on the *Friedrich der Grosse* moderate crew-members had to hide Commander Oldekop, Reuter's chief of staff, to prevent his being lynched.

On 20 January there was a putsch aboard the seething flagship, organised by the Soldiers' Council. It voted to dismiss the captain and the officers, and elected a coxswain to take command. Dominik felt it was impossible to go on flying his flag on the *Friedrich der*

Grosse and decided to go back to his own ship, the *Bayern*. The Soldiers' Council protested vehemently, saying that the disorder on the flagship at the time would be as nothing compared with what they thought would happen if he left. Dominik found such a further deterioration unimaginable; short of an outbreak of killing he doubted whether the situation could get worse than it already was, and he said his decision to move stood. The Soldiers' Council then undertook to restore some semblance of order if he would agree to stay until Reuter got back. Dominik was thus able to gain a small but significant victory before Reuter returned. He had shown that a firm stand by the senior officers on a point of fundamental principle could frustrate the revolutionaries and that the German yearning for order had not been entirely extinguished among the crews. The lesson was not forgotten, but a great deal of danger for the officers still lay ahead.

The struggle for power had begun before the ships set off for internment, with several arriving in British waters without officers aboard. While the Soldiers' Councils rejected some soon afterwards, they confirmed in their commands officers who had been popular or at least not unpopular before the revolt. Reuter had been able to send home many rejected officers in the main repatriation exercise of the first half of December, without appearing to concede anything to the Councils against the background of a 75 per cent overall reduction in manning. The British refusal to allow the officers, including two Commodores (Harder and Tägert), to travel home separately condemned them to discomfort, insult and humiliation on the transports, on which rebellious sailors set up temporary councils for the two-day voyages to Wilhelmshaven. Many attempts were made to eject officers from their cabins as the crews went home in conditions of indescribable squalor.

Although Reuter had reduced the number of officers to a minimum through the repatriation, the same did not apply to the agitators, most of whom opted to stay. The Admiral subsequently took every opportunity to reduce their numbers whenever an excuse, such as demobilisation of certain age-groups or men with a certain length of service, offered itself. Although his principal reason for keeping the number of replacements down to a minimum was to avoid any increase in agitators, even the Councils came to agree that those who had brought the ships across might as well sit out internment and that there was no point in making thousands of other men go through the painful process of adjustment; and such co-operation and teamwork as existed would obviously be ruined by crew changes. A British order of 22 February, motivated by fear of further infiltration by agitators, helped Reuter by imposing a complete ban on replacements except where they were indispensable on grounds of special skills.

Apart from his own staff officers, Reuter had a small number of specialists on hand, including a chief surgeon, a fleet paymaster, a fleet engineer and a legal adviser. Engineer officers were especially unpopular with the men, probably because they inevitably made the greatest demands for labour in supervising maintenance. The post of engineering adviser on the flagship he gave to Engineer Müller of the *Friedrich der Grosse*. The Soldiers' Council objected strongly to Müller's presence and only came grudgingly to accept it when a contract was drawn up in which Reuter listed, and thereby limited, his powers. There was an acute shortage of engineering expertise throughout internment, but Müller did a good job.

The confrontation with the Councils continued unabated after Reuter's return on 25 January. One of the major battles was fought over the question of discipline. Reuter had very few cards to play. With all the crews confined to their ships by the British they were effectively prisoners anyway, so the idea of detention, or confinement within confinement, was little more than a joke. Putting men in cells was hardly an option open to him given the general rebelliousness. Fines deducted from pay were possible but not a formidable deterrent in the circumstances. When a serious offender was sent home it was very easy for him to evade trial and punishment; if he were handed over to the British for punishment he at least escaped from internment and could look forward to a leisurely train journey through the Scottish Highlands to a naval prison at Perth. As part of his general policy of keeping British intervention to a minimum by avoiding giving them excuses, Reuter was not in any case disposed to fall back on their authority in support of his own precarious position. He was also conscious of a lack of information about legal amendments in Germany after the change of regime there. The strict limits on movement among the ships made it well nigh impossible for a naval court to assemble. Faced with these formidable difficulties, Reuter decided that the best form of defence was attack. He ordered the exclusion of the Soldiers' Councils from any participation in the administration of justice and questions of internal discipline. The revolutionaries were furious and demanded the application of the Emergency Law of the Republic of Oldenburg as the basis for law and order on the ships. Reuter, relying on the Reich Government decree of 19 January limiting the role of the Councils and restoring some authority to officers, stood firm. He won. Eventually the traditional naval disciplinary code of the Kaiser's days, modified to take account of the special circumstances of internment, was accepted as the basis for discipline (which did not mean that it was observed particularly closely). Over-riding further protests, Reuter insisted that only the captain had the right of ordering an arrest aboard ship, except where an offender was caught in the act by another person in authority. Officers were encouraged to turn a blind eye to as many infringements as possible to avoid unnecessary or over-frequent confrontations. Fines became the main punishment at first; later it became possible to put some more serious offenders in cells. It was a more important moral victory on the endless road to the reassertion of proper authority than had been the decision that officers had sole responsibility for seamanship on the voyage into internment.

But the struggle was far from over. Almost every day there was some new row between the officers and the Councils. Fresh waves of unrest broke out on the *Friedrich der Grosse* after a Leading Torpedo Mechanic smuggled himself aboard the flagship on 10 February from the destroyer *B98*, which had come from Wilhelmshaven with despatches, supplies and replacement personnel. The principal agitator on the flagship however, Leading Seaman Grahlmann, was due for demobilisation on 31 March, and Reuter exhibited no reluctance towards his departure, even though he changed flagship at the time. One can only wonder at the inexhaustible sense of duty and the sheer determination of Admiral von Reuter, the naval answer to King Canute. Considering that he had every right to resign after leading the ships into internment, it is astonishing that he ever came back from leave. No doubt his colleagues were instrumental in persuading him to do so; but it is clear that he had more than enough strength of character to resist mere blandishments had he chosen to. He certainly came back with a free hand to act as he thought fit, and he was not slow to use it. No commander has ever been placed in a position comparable with his before or since. His eventual decision to scuttle brought him a certain glory, but that was not to be foreseen. He did a lot for his 1,312 Marks a month.

The climax, though not the end, of his confrontation with the revolutionaries came when he took the bold decision to dismiss two members of the Supreme Soldiers' Council of the Internment Formation, a frontal assault on their presumptions of authority. The opportunity came when the two men, stokers called Müller and Willershausen, the two most extreme members of the Council, were caught in flagrant breach of British orders and also of the agreement between the officers and the Council that they should be seen to help preserve the ships for Germany. The two agitators joined a party of German sailors helping to unload a supply ship so that they could smuggle themselves aboard it for revolutionary purposes. There was, needless to say, uproar over Reuter's decision to dismiss them. His reference to the Government Decree of 19 January was rebuffed with the same government's acceptance of the role of the Soldiers' Councils in response to their demands of 24 January. Reuter admits in his book that he was exceeding his authority in dealing with the Councils but 'the Senior German Officer in Foreign Waters surely must have the right to expel incompetent delegates whose conduct endangers friendly relations with the "host" state'. He decided on a tactical withdrawal for the time being and agreed that the matter should be referred to the Reich Naval Office for a final decision. The dispute had the not entirely unsatisfactory effect of splitting the crews. The ever-volatile crew of the *Friedrich der Grosse* went on strike, no easy task for men devoted to doing as little constructive work as possible anyway. There were calls for the dismissal of Reuter and he was subjected to personal insults. Then the British took a hand in the controversy, which went on for a month until its dramatic conclusion, calling for the two men to be handed over to them to restore some semblance of peace. Reuter refused and the Royal Navy did not press the matter: but Reuter took advantage of the British interest by presenting the two men with a choice: either stop agitating and return home or face being handed over to the British. Müller and Willershausen opted to go home and Reuter knew he had won. Even without a decision in his favour by the Reich Naval Office, he was to get rid of the two agitators. In the event, Berlin ruled in favour of Reuter in a radio signal about a week later, and said the two revolutionaries were to be repatriated on the first available ship. The letter which followed the message after a distinctly un-urgent interval of three months 'reminded' the Admiral that he had no power to dismiss members of the Supreme Soldiers' Council.

The Admiral, meanwhile, was still determined to find a quieter flagship. Shortly after his return he went to the British flagship after dark, an unprecedented event, to raise the possibility of a move to the battlecruiser *Hindenburg*. The crisis on the *Friedrich der Grosse* persuaded him to put off the decision, and he told the British he would not be moving for the time being. Thirty agitators were sent home on a supply ship on 17 February, which helped. Then he considered a move to the *Baden*, a former flagship, but thought better of that when he remembered how the ship had been overrun by women in the outbreak of the revolution. He considered the *Kaiser*, still the most orderly of the battleships, and then one of the largest destroyers (rejected on grounds of discomfort, lack of facilities and small size). For a while he felt that any move might be interpreted as a surrender by both the British and the Soldiers' Councils, but when in the middle of March the crew of the *Emden* went so far as to offer him refuge, his mind was made up.

The next supply ship, once again the destroyer *B98*, was due on 23 March, returning to Germany the next day. Reuter decided to put the two revolutionaries aboard and to transfer his flag to the light cruiser on the same morning, of 24 March. The *B98* arrived, and the two agitators suddenly refused to go aboard. The British, who had been following events with understandable interest, took a hand and gave them two hours to get aboard when it was learned that they wanted to break their word. Playing for high stakes as he was, Reuter did not shrink from explaining his difficulty to the British. A Royal Navy destroyer and an armed steamer, guns and torpedo tubes cleared for action and trained on the *Friedrich der Grosse*, moved into position. The two troublemakers bowed to the inevitable and boarded the *B98*. When they left, two other more moderate members of the Supreme Soldiers' Council resigned. Reuter, showing no outward sign of satisfaction, dutifully called an election to fill the vacancies on the Council, which was duly held in April, but he never again had quite the same degree of difficulty with the revolutionary representatives, who were divided among themselves and who were steadily losing the support of both the remaining extremists and also of the crews. The discontent simmered on, but the last three months of internment left Reuter free to sleep easy in his bunk most nights. The officers had the stronger hand, provided they played it with care.

The ensuing move to the *Emden* drew objections from the discomfited revolutionaries because it meant the Admiral would be out of reach; the announcement of the move had a bombshell effect. The British provided three drifters for the move and placed armed Royal Marines on the *Friedrich der Grosse*, but there was no trouble. The Rear-Admiral in Charge, German Ships at Scapa, as the British now referred to him, took up with relief a distinctly less stressful existence. The British understood his difficulties but still thought the move to a humble light cruiser rather bad form. His earlier idea of moving to a battlecruiser did not win their approval either. A fleet commander belonged on a battleship or nowhere. Reuter explained, both to the British and his own crews, that he had acquired the taste for light cruisers in his days as a reconnaissance group chief; that the *Emden* was better placed because it could make direct signals to the British flagship (Cava was not in the way); and that the atmosphere on the *Friedrich der Grosse* had made things difficult for him.

More than six weeks of relative calm followed the events of 24 March and the monotonous days of isolation wore on. Men supplemented their diet by angling with improvised rod and line from the deck. Now and again somebody would still make a political speech, but the shortage of information from home robbed such oratory of immediacy and direct relevance. The new Supreme Council elected in April proved distinctly more moderate than its predecessor. The circumstances were ideal for the growth of a flourishing rumour industry. Reuter's change of flagship unleashed a great deal of speculation: were the capital ships to be towed away and sunk, or divided up among the Allies or broken up for scrap? Having weathered a northern winter of some severity, the neglected ships had lost the last remnants of external dignity. One or two crews spontaneously decided to fill their time by repainting their ships. Even the *Friedrich der Grosse* got a new coat of battleship grey thanks to the dedication of a boatswain's mate who organised it. Reuter was pleased and ordered a general external clean-up of the whole fleet. This also sparked off rumours of impending changes. The removal by the British of radio receivers after the putsch on the *Friedrich der Grosse* made news a highly-prized commodity. Such German newspapers as reached the crews tended to be a good fortnight out of date, and there was the usual proportion of those who could not resist the temptation to use their imaginations to fill the gap. Revolutionaries also regularly used false information as a tactic in their agitation. The freshest and most reliable information reaching the interned Germans was extracted from copies of British newspapers, usually *The Times*, which the Royal Navy was quite happy to let them read. The papers were invariably four days old, a fact which has great significance in the story of the scuttling. Every now and again the British press would remember the presence in Scapa Flow of the enemy navy. The *Daily Mail* reported that internees had been sneaking ashore in boats and stealing sheep to supplement their diet. There was no truth in the story, which a British imagination had extrapolated from a Royal Navy order that all small boats on the islands overlooking the Germans should be hauled up on shore for security reasons. The British were always sensitive to the possibility of the Germans using small boats for nefarious purposes; so much so that when Reuter sought permission to lower boats to help with the great repainting, he was turned down flat. He admitted to a mixed motive: if wooden boats are kept out of water too long, their planking warps and they leak badly when relaunched. With the possibility of scuttling always at the back of his mind, he wanted to ensure that the boats remained usable.

Reuter continued to concern himself with domestic matters, sending a constant stream of requests for supplies to Wilhelmshaven, notably for cigarettes and the elusive dentist. Wilhelmshaven replied with such urgent requests as the prompt return of empty beer barrels. He also wanted a say in the future of the German navy after peace was signed. On 23 April, for example, he sent a message to the Reich Naval Office urging them not to make a final decision about changes in naval uniform until they had received his own proposals in the form of a letter he would send as soon as he could. He also voiced the worries of his officers about their future after internment came to an end. They had put their concerns on paper, pointing out that unlike their colleagues ashore they had no opportunity to prepare themselves for civilian life because they were completely isolated. They drew up a list of requests not dissimilar to the document drawn up by the Soldiers' Councils on behalf of the crews in January. Reuter passed them on in March and the Naval Office agreed to virtually all of them. They wanted posts kept open for those who wished to go on serving; up to three months' leave plus four and a half months' pay over and above the gratuity for those leaving, and the like.

In March, the British broached the subject of a second large-scale reduction in manning levels. They wanted the crews to be reduced to the kind of levels adopted in the Royal Navy for ships of the reserve, maintained in port by skeleton crews rather less than half the size of the German caretaker parties in the interned ships. Reuter resisted the idea on grounds of safety and comfort. The ships after all were not in port and had no contact with shore facilities. He counter-proposed a much smaller reduction of from five to twenty men, depending on the class of ship. The British did not press the matter and nothing was done. Reuter was suspicious. Were the British concerned to eliminate agitation or were they trying to make it easier to seize the ships? The large-scale reductions they proposed would have the effect of immobilising the ships. Maintenance was already rudimentary and it was doubtful whether many now could live up to his requirement of being able to get up steam for ten knots in case of a move elsewhere. A drastic cut would reduce them to helpless hulks. Reuter had managed to send home about 150 agitators in the wake of his victory over the Supreme Soldiers' Council. There the matter rested for two months.

Disruption broke out again, however, after the first week in May when word of the Allied naval peace terms reached Scapa Flow. These reduced the German navy to little more than a token force (see below) and caused a tidal wave of speculation and renewed agitation among the crews. One of Reuter's strengths was his ability to combine resolve with the capacity to change his mind: he always accepted that circumstances alter cases. So he began to reconsider the question of another large reduction in the crews in the light of the new disorder. He cast aside, according to his book, his last doubts about the adverse effect on the seaworthiness of the ships when he uncovered towards the end of the month a Soldiers' Council plot to call a general strike. The aim was to make the German Government seek the release of the crews, but Reuter recognised that this would provide the British with an excuse for seizure of what remained German property, at any rate for the time being. The loss of seaworthiness, which could eventually be restored, now seemed a lesser evil than the likelihood of the loss of the ships, which most probably could not. Oldekop, still Chief of Staff, was thinking on similar lines and suggested sending home all crew-members who wanted to go so long as enough remained aboard to prevent damage and keep essential services running. Reuter agreed and resolved that 'the reduction would be carried out with severity sufficient for the sinking to be completed with efficiency. The state of readiness for sea was dropped.' Scuttling was now at the forefront of his mind. He obtained ready British consent to reduce the crews to Royal Navy caretaker levels, as follows:

Battlecruiser:	75 (hitherto 200)
Battleship:	60 (hitherto 175)

Light cruiser:	30 (hitherto 80)
Destroyer:	as their Leader saw fit, previously 20

Conditions were as good as they were ever likely to be. Midsummer was approaching and the evenings were already very light in the northerly latitude of Scapa Flow. The skeleton crews could use the naturally lit parts of the ships in reasonable weather, reducing the need for power for lighting and heating. Coal, oil and water shortages would disappear and provisions would be plentiful. Boilers could be shut down and care and maintenance dispensed with. Only cooks and a few hands for general duties were needed. Commander Cordes decided he needed no more than 400 men to look after his fifty destroyers. Many of them, moored in pairs as they were, were left unoccupied and rat-infested, subject to occasional safety checks from the occupied ships. All this meant that the interned personnel could be cut to about 1,700. Two transports, the SS *Badenia* and the SS *Schleswig*, eventually arrived on 17 June, and on the following day 2,700 men were embarked for repatriation, just three days before the scuttling.

CHAPTER EIGHT

Reuter's Decision

APART FROM THE understandable unease caused by the news of the peace terms, the final weeks of internment passed off with only one major incident – on the anniversary of the opening day of the two-day Battle of Jutland, known to the Germans as the Battle of the Skagerrak, on 31 May. Even May Day, declared a public holiday by the German government for the first time in 1919, was not marked by any special celebration on the interned ships. In an Order of the Day of 30 April, Reuter said he felt sure the crews would share his view that a celebration would be inappropriate. The peace terms were imminent, and 'these [coming] days could bring us in particular a specially bitter decision'. A new diversion arrived to counteract the all-pervading 'Scapa Flow syndrome', also known by the crews as armour-plate disease, in the form of cinema films which, Reuter wrote, were 'received with rapture'. But the novelty soon wore off once all the films available had been swapped and shown round the fleet.

But as the Skagerrak anniversary approached, a rumour (again) went round the interned fleet to the effect that the British were planning to seize the ships on the day *they* celebrated the battle, on 1 June. The Soldiers' Councils believed the story and issued orders for the destruction of papers should the British board the ships and hoist their flag on them. Reuter and his staff did not give credence to the rumour, but they were concerned lest the actions of the German crews on the anniversary might provoke the British to intervene. The captains were quietly told to be prepared to sink the ships if the attempt was made, but not to inform the men of the plan. The crews were told that they could celebrate the occasion, but below deck. The destroyer men in particular, recognised by the British as well as Reuter as having the highest morale and the best relations between officers and men, had other ideas. They rigged up chains of light bulbs to make slogans such as 'Victors of the Skagerrak, 1916'. Many ships were dressed overall with bunting. There was a flurry of activity by the patrolling drifters, but nothing more. Reuter recalls in his report how the many red flags which were also produced were reflected in the calm water of Scapa Flow in the evening light. The Germans detected no sign the next day of any British festivities to mark the battle. The last party was over. Meanwhile at Versailles the interned German ships were proving one of the knottiest problems for the Allies as they worked on the draft Peace Treaty. While all were agreed that the vessels – and others – should be taken from Germany as part of war reparations what was then to happen to them remained an unsettled issue which led to protracted wrangling on the Allied Naval Council and among the statesmen and diplomats.

The French took an early hand in the debate. As early as the beginning of December 1918 they asked for a quarter of the interned ships. They also asked that some of them should be sent to French naval bases so that they could be shown as tangible fruits of a hard-won victory to those who would otherwise never see what they had helped to achieve, just as people in northern France could see the captured guns and other spoils from the Western Front. The French, and the Italians too, also took the view that they were entitled to a significant share of the naval booty, because they had neglected their navies during the war in order to concentrate their resources on their armies. The British, naturally enough, were not keen to see the ships used to strengthen other navies. Since they retained naval supremacy, and they had no use for extra ships themselves, especially when they were busy with demobilisation in the Royal Navy, reorganisation and reduction of the fleet and placing ships in reserve as rapidly as they could. However the Americans, who were soon to begin the next challenge to British naval predominance, which was destined to succeed, would not have minded if other navies had been expanded at the expense of the British. Other options under consideration included a vague proposal for using the German ships as an international peace force, and the straightforward proposition that they should be taken out into the open sea and sunk. According to the American naval historian Arthur J. Marder, in his book *From the Dreadnought to Scapa Flow*, this was Lloyd George's private view in April 1919 as the disagreements wore on. Apparently he would have liked to see the ships towed to the middle of the Atlantic, where, with the national anthems playing, they could have been ostentatiously sunk. President Wilson, however, wrote to a Congressman that this idea 'strikes me as the advice of people who do not know what else to do'.

In public Lloyd George told the British Parliament (at the end of February 1919) that the ships would be counted as part of Germany's war reparations. Lord Lytton told the House of Lords that the ships would not be sunk unless all the Powers concerned unanimously accepted this as the best solution. He went so far as to add that, meanwhile, the ships would not become part of any fleet in the world and would not be passed on to the navy of any other nation. The French government, on the other hand, said that the ships would not be destroyed, except for a few 'to teach the Germans a lesson'.

The Germans were not unaware of the deadlock among the Allies on this and other issues. The Armistice had been repeatedly prolonged while various points were thrashed out and it took the Allies six months to produce the final draft of the Treaty of Versailles. But the Germans had no illusions about getting their navy back. Reuter accuses the government of resignation and of failing even to try to salvage something of the fleet, but it is better seen as realism. The surrender of the submarine fleet in its entirety and forever had already happened as a condition for the Armistice, not the Peace, and in the end more than 200 U-boats were handed over (fourteen others scuttled themselves). It would have been foolish to expect the retention of anything more than a token surface fleet. Besides, the Germans had nothing to bargain with, as their negotiators realised, even if naval officers did not.

The peace terms were finally ready on 6 May and were formally handed to the Germans on the following day. The draft treaty came in fifteen sections incorporating 440 articles. The elements which concern us here, the naval conditions, were as harsh as the German negotiators feared. All the interned ships were to be handed over, together with another fifty-two torpedoboats. All warships under construction were to be dismantled. Germany was to be left with a navy of no more than 15,000 men and 1,500 officers. She was to be allowed no more than six battleships of a maximum of 10,000 tons, and six cruisers of up to 6,000 tons, twelve small destroyers of up to 800 tons and twelve small torpedoboats of up to 200 tons. The navy was not allowed submarines or aircraft. The military installations and harbour at Heligoland were to be dismantled.

The news, when it reached Scapa Flow on 11 May, the usual four days late because of the delayed newspapers, was naturally a shattering blow. Even the rebellious elements were depressed because they too had hoped that something would be saved by co-operation with internment. 'For a couple of days it lay like lead on the minds of the men,' Reuter wrote in his book. After a while the crews began to cheer up; at least they could now see an early end to isolation in their hated anchorage.

The Admiral was not given any information or text by the British and had to wait some days before he got the chance to study a German text which he may have acquired (he does not say) from a supply ship, perhaps in the form of extended quotation in a

German newspaper. He says he received no word from Germany either of information about the terms, or any kind of order.

He began to consider the possibilities open to the German government and his own options. 'It was quite clear to me that I should be left entirely to my own devices. I had refused orders and instructions because I alone and nobody at home could assess and appreciate the situation in the Internment Formation.' The German government could reject the terms, negotiate improvements or accept them. A refusal meant a renewal of hostilities and a British attempt to seize the ships. It would be treason to let them. As far as he and his fellow officers were concerned, the standing order that no ship was to be allowed to fall into the hands of an enemy under any circumstances, laid down by the Kaiser as 'All-highest' War Lord, still applied. Under these conditions the ships remained German property and there would be no option but to scuttle them because they were defenceless.

Negotiation might mean the use of the ships as a bargaining counter in exchange for concessions elsewhere. In such event, the interned fleet would have achieved a German objective and would have to be handed over to honour a contract, however distasteful this might be to the officers. Reuter viewed this possibility as the most difficult of all and the silence from Germany about government intentions cost him a lot of sleep. He could not believe that the ships would be offered for sale. Was the government secretly committed to this and refusing to reveal it? The officers would in honour have to oppose it. There was the example of the fate of the submarine fleet and the abiding fear of the German naval officer that when the chips were down the army always came first. Would the navy be sacrificed to save the army?

The third possibility was unconditional acceptance of surrendering the ships. Such German press reports as he had seen convinced Reuter that this was inconceivable. It could only be contemplated on receipt of a direct order from Germany. Reuter realised he was beset with problems. The crews were unreliable (a problem he solved by deciding to pare them to the bone when he heard the peace terms). Communications with Germany were minimal, slow and subject to enemy vigilance. Finally, there was the delicate question of whether the peace terms, once accepted by Germany in whatever form, come into effect on signature or on ratification. His greatest fear was that the British would act pre-emptively and seize the ships on signature without waiting for ratification, and he decided on his own responsibility that such action on their part could be met in only one way – scuttling.

Naval officers on both sides at Scapa Flow had devoted much thought to the possibility of scuttling from the very beginning of internment, the Germans to how it could be achieved under the eyes of the enemy should it become necessary and the British to how it might be forestalled or minimised. The instinct of the thinking naval officer on either side led him to conclude, if only subconsciously, that there would surely be no other outcome given Germany's hopeless position at the peace talks, whatever plans the politicians and negotiators might draw up. The British could hardly have failed to realise what kind of a man Reuter was – a man after their own hearts even if they could not admit it. They knew what they would have done themselves, and Reuter knew they knew.

The British were aware that they were at a complete disadvantage in the event of a German decision to scuttle. They were precluded from putting armed guards aboard the hostage fleet because of its interned status, and even had they felt free to do so the sabotage necessary for sinking could, if properly organised, be completed in a few minutes. Once the seacocks were opened, sinking could not be prevented and the only hope was to react quickly enough to be able to beach the ship before it foundered, which could be done only by towing, as there would be no power aboard. Even so many of the ships would go to the bottom.

Scuttling was anticipated by the Royal Navy from the very beginning, as the standing orders to the guarding force showed. Successive Senior Naval Officers Afloat at Scapa drew up orders for seizure of the ships to prevent them being sunk. Yet, according to Marder, Vice-Admiral Sir Sydney Fremantle, who had taken over responsibility for surveillance in May, was surprised at not receiving instructions about the disposal of the ships and their crews as the deadline for the German decision on the peace terms drew near. Thus on 16 June he completed detailed orders for seizing them at noon on 21 June, the deadline given to the Germans in an Allied ultimatum of 16 June, and sent them to his C-in-C, Admiral Madden of the Atlantic Fleet, for approval. He had drafted them on 21 May and revised them on 29 May. A British guard was to be placed on each German ship and the crews transferred either to British ships or concentrated on a few of their own. On the hoisting of the signal 'M Y' with international code flags, the armed parties, sixty men for each capital ship and twenty-nine for a light cruiser, were to board and go below to prevent sabotage. The German crew was to be evacuated and one petty officer and four ratings, including a signaller, were to stay aboard and hoist a pennant to show their work was completed. Eight destroyers were to be under way round the anchorage and at action stations, and a staff officer and an interpreter were detailed to go from Fremantle's flagship, the *Revenge*, to the *Emden* to fetch Reuter. Looting, pilfering and 'collection of souvenirs' were strictly forbidden on pain of severe punishment. The German crews were to be landed in the Cromarty Firth, and 'it is to be remembered that their status will have become that of prisoners of war'. On 19 June Madden approved this admirably thought-out plan, which seemed to cater for every eventuality except one – that the Germans might act first. Madden was told by telegram on 17 June that the deadline for signature of the peace treaty had been extended from 21 to 23 June. According to Marder, Madden himself later regretted not passing this on to Fremantle at once. On 20 June Allied representatives authorised the seizure of the interned ships at the moment the Armistice, extended by then to 7 p.m. on 23 June, expired. Fremantle, however, had decided to take his squadron of battleships to sea for exercises on the morning of 21 June and to return on the afternoon of the 23rd for the execution of this plan that evening if necessary. Amazingly, Reuter (the news agency, not the Admiral) was able to report on 21 June that the Allies planned to seize the interned ships two days later unless the Germans agreed to sign the Treaty. In keeping with their policy of isolating the German ships as completely as possible, the British told Reuter nothing about the last moves before peace was concluded. He was not notified of the 21 June deadline or of its extension to 23 June. The information on which he says he acted was thus unofficial, accurate, but out of date because of the bewildering speed of political changes in Germany and in the negotiation in the last hours of the Armistice.

The British were absolutely right in anticipating a decision to scuttle. The Germans had begun to think about it even before they left for the Firth of Forth. While they were in Wilhelmshaven waiting to depart some officers had spoken in favour of scuttling in the North Sea as soon as they were clear of the coast. The thought was always in Reuter's mind; he must have discussed it while on leave, and he says he raised it with Oldekop at the end of January about a week after his return from Germany. The two officers debated it a second time at the end of March after the crisis with the Supreme Soldiers' Council and Reuter's move to the *Emden*. By then hints had already begun to appear in the British press of the nature of the peace terms, confirmed by a report from the German negotiators at the conference which deduced that the Allies wanted the surrender of the ships, and the German government was unlikely to resist. In his book Reuter said: 'I have not been able to rid myself of the idea that the [German] Government had already given up the Interned Formation at the time of the Armistice; their conduct at the Armistice negotiations and their acquiescence in the detention of the German ships at Scapa Flow all point to this. It therefore appeared to me that the preservation of our ownership of the ships was put entirely in the hands of the Commander of the Interned Formation.' For their part, the British, he thought, wanted the ships to become unserviceable, making them as unattractive and immobile as possible. His suspicions of British motives had been strengthened when they raised the question of reducing the crews to skeleton level. In London the press suggested that Britain was opposed to a share-out and called for the ships to be sunk. But Reuter felt he was honour-bound to wait for full details of the peace terms before he made a decision. The notion of the honour of the German navy played an increasing part in his deliberations.

Active planning for scuttling began in the middle of May, when the Germans got details, the usual four days late, of the peace terms. There were a number of difficulties. Many officers began to urge Reuter to prepare to scuttle and he was obliged to rebuke them for voicing his own sentiments, to prevent the idea spreading round the rumour-hungry fleet. Nor could practical preparations be undertaken until the last possible moment in case of a leak to the British. Then there was the fact that German warships were designed to resist sinking. Many of the destroyers for example had hulls divided into twelve watertight compartments and were designed to stay afloat even if all but two of these were flooded. At Jutland one battlecruiser, the *Seydlitz*, had received more than twenty direct hits from British heavy guns and managed to get home in twenty hours.

Unaware of the exact nature of the cogitations of their Admiral, the officers and men of the destroyers drew up their own plans to scuttle. Their leader, Commander Cordes, chaired a conference on the *S138* on 13 May which led to a series of active steps. Each watertight compartment had a water-valve for fire-fighting purposes which could be operated from above deck as well as below. The connecting rods which enabled the flood valves to be operated from above were therefore removed so that a boarding party would be unable to turn them off. Officers were assigned to sleep in the engine compartments of unoccupied destroyers. Hammers and axes were left in boiler rooms and engine compartments to smash pipes carrying water, and the screws holding the condenser-covers in position were taken out (every time power was needed they had to be put in again). There were no explosives that could be attached to the seacocks, and officers were concerned in case six months' accumulation of seaweed on the hulls might stop some flood valves from functioning properly. Watertight doors were left open and holes bored through bulkheads to facilitate the rapid spread of incoming water, and portholes were loosened. So tight was security that Reuter himself did not find out about the destroyer men's independent preparations until the beginning of June. He summonded Cordes and told him of his plans to scuttle the entire fleet, warning him of the danger of a leak to the British.

All the officers had in fact guessed which way the wind was blowing by the end of May because Reuter had then ordered his captains to be ready to scuttle if the British moved in on the anniversary of the Battle of Jutland; the crews, he said, were not to be alerted in advance, and a careful watch against surprise by the British was imposed on officers while the men celebrated.

Tension truly began to mount for Reuter and therefore the staff, who were in on the plan to scuttle from the beginning of June, when they learned of the rejection by the government of the peace terms in a speech to the Reichstag by Chancellor Scheidemann on 29 May. The news reached them via officers on a despatch boat from Wilhelmshaven. At the same time hints were picked up from Royal Navy personnel about the existence of a plan to seize the ships on the signature, if any, of the Treaty, rather than on its ratification. Reuter was now clear in his own mind that he would scuttle if the British attempted to seize the ships at any time before ratification or if the German government finally rejected the peace terms, which implied a renewal of hostilities. Only a decision by Germany to accept the terms, made known officially to him, would now stay his hand. For the success of the plan a great deal depended on the completion of the last crew-reduction scheduled for the middle of June; Reuter hoped desperately that the transports coming to take them home would not be delayed.

Reuter insisted to the end of his days that he did all this on his own initiative. No evidence to the contrary has been found in the German Archives or, more significantly perhaps, in the British Public Records. The British did everything in their power to prove that he had acted on orders from Germany. There is no proof that he did; there is no proof (how could there be?) but only a strong likelihood that he did not. Only one official communication from Germany to Reuter about the implications for the navy of the peace terms survives from the period between the announcement of the conditions and the scuttling of the High Seas Fleet. It is a letter from Vice-Admiral von Trotha, State Secretary (head) of the Reich Naval Office, dated 9 May, which was smuggled to Reuter by *Kapitänleutnant* Stapenhorst of the light cruiser *Königsberg* when that ship brought despatches and supplies to Scapa Flow on 18 May. The key passage in this letter translates in full as follows:

> Your Honour may be certain that it will be the obvious duty of our naval negotiators in Versailles to defend the fate of the Interned Formation by all (possible) means and to bring about a solution in keeping with our traditions and Germany's unambiguous rights (or 'law' – the German could mean either). In this regard, priority will be given to the condition that the Formation should remain German, that its fate, whatever form it takes under the pressure of the political situation, will not be determined without our participation and will be executed by our own selves, and that a handover to the enemy remains excluded. We must hope that these just

> proposals will be successfully asserted in the framework of our general policy position on the peace issue.

The letter concludes with the hope that Germany could keep the ships, an assurance that the internees had not been forgotten and a reminder that the fate of the fleet depended on the success of the Versailles negotiators. The second sentence of the above was taken out of its context by the British after the scuttling as proof that Reuter did not jump but was pushed. They based their entire case on it, which shows they had no case at all. The letter was obviously meant as a reassurance, written as it was three days after the peace terms became known. All it does is to outline a negotiating position for the next round of talks, and it is expressed in such a way that it becomes clear when it is read in its proper context that Trotha is doing little more than expressing a pious hope, as any naval officer would, that the German negotiators would put up a successful fight for the ships. Reuter did not destroy the letter; he put it in the safe of the *Emden* and forgot about it.

Reuter drew up a detailed order for scuttling the ships on 17 June (see Appendix II). It gave no date of execution, and was addressed to all captains and the Leader of Torpedoboats, to be delivered by officer only. The commanders were instructed to prepare their ships for sinking on the lines already followed on the destroyers, and to have lifeboats ready for evacuation, complete with a white flag to deter the British guard from firing at them. Scuttling was to take place if the British occupied the ships by force, or in any case on receipt of a special order from Reuter. Two coded signals, one for immediate sinking and the other for scuttling at a later time, were included. Maximum alertness for signals or unusual British activity was to be kept from 18 June, by officers. The order concluded:

> It is my intention to scuttle the ships only if the enemy should attempt to place himself in possession of them without the

> consent of our Government. Should our Government concur with the surrender of our ships in the peace conditions, then the ships will be given up, to the lasting shame of those who put us in this situation.

The only change detected by the Germans in British behaviour towards them once the peace terms became known was a sharp increase in vigilance on the part of the censors, which meant even longer delays in the delivery of post. This made the Germans nervous and increased the anxiety of those concerned in planning the sinking. They could in their isolated condition never be certain that their main advantage of being in possession of their ships without guards aboard might not be offset by the British advantage of being four days ahead in their knowledge of political developments in connection with the decision on peace.

There was great relief aboard the *Emden* when Reuter and Oldekop saw the two transports arrive on the 15th and the 17th to take off the men who were to be repatriated. They were embarked at high speed, and as they left on the morning of the 18th, British drifters obligingly and innocently went round the interned ships delivering copies of Reuter's order, an irony he recalled with relish. It had always been a simple matter to conceal communications from one ship to another during internment, especially when supplies from Germany were being distributed by the patrol-vessels, just as it was easy to send and receive papers via the warships from Germany which regularly acted as despatch boats to Scapa. Officers and Soldiers' Councils alike made liberal use of both facilities. Important messages could also be passed on orally.

On 20 June Reuter learned that the secret of the plan to scuttle was out on at least two of the capital ships. The captain of a battle-cruiser found that members of the crew were dogging his footsteps as he went round the ship to make preparations. Petty officers aboard a battleship guessed what was going on by observing the actions of the officers. What is surprising is that the secret was not guessed all round the fleet or by the British at Scapa. But many German sailors wanted nothing to do with the plan for scuttling once it became generally known. Accepting the inevitable, Reuter issued a second order giving captains discretion to tell their crews. Some did, some kept silent until the scuttling of their ships had actually started. Again the British unknowingly distributed the new order that afternoon. The business of drafting and distributing the order kept Reuter busy all day, so that it was only in the evening that he learned from *The Times* of 16 June (four days old as usual) that the Allies had given Germany a new ultimatum to accept the peace terms by noon on 21 June or face a renewal of hostilities. He decided to order the scuttling the next morning, 21 June.

On 20 June, however, the anti-treaty Scheidemann resigned and was replaced by Bauer. On the same day the Reichstag voted by 237 votes to 138 in favour of accepting the terms. The Armistice was extended first to the morning, then to the evening of 23 June (the Germans eventually signed on the 28th). Reuter insists he knew nothing of all this until the evening after the fleet went to the bottom.

Although the Germans were deliberately kept in ignorance of the political developments by the British, the German Archives reveal that at least a few of the internees knew of the last extension of the armistice before Reuter issued the order to scuttle.

In the Archives there is a well-bound blue volume of impeccable typescript by Commander (retired) Johann Heinemann, embellished with a romantic photograph of a painting showing a light cruiser ploughing through a rough sea. It was the story of the *Cöln* by its captain, a *Kapitänleutnant* in 1919, and it verged on the length of a novel. Near the end there is a reference to the *Cöln*'s radio log. She should not of course have had one, given that both transmitting and receiving had been banned by the British, and the necessary radio parts removed to enforce the order. But the report said that a signaller had given Heinemann a garbled report of a broadcast he had picked up, faintly and with much interference, on a small, home-made radio receiver. Heinemann wrote:

> I thought I could deduce from this that the expiry of the Armistice had been deferred by two days, i.e. to the following Monday. Thus the critical day would not be the 21st, which fell on a Saturday. I considered whether I should not tell the Admiral this. In the end I let the idea drop. After all, on the *Emden* too they would have knocked a receiver together – sheer conjecture on my part – and I would have nothing new to report to the Admiral.

On the contrary!

By the evening of 20 June, Reuter's sense of honour was fully engaged as he decided to scuttle. He inferred from his struggle with the English of *The Times* that the Germans might still go so far as to surrender all the ships in their possession, not only the interned ones, in exchange for a financial easement. What a humiliation for the Fatherland, he thought. He had been led to believe that the government would fight for the return of the interned ships; now they seemed to be offering even more than they were being asked to surrender. He reached for pen and paper and began to draft a protest on behalf of the officers, to be sent to Germany via the British flagship for transmission the next morning. It was never sent. First thing the next morning, the British battle squadron of five battleships and attendant smaller vessels steamed out of Scapa Flow for exercises. The scene was set, the decision taken. The ships of the German High Seas Fleet were doomed.

CHAPTER NINE

'Paragraph Eleven. Confirm.'

REAR-ADMIRAL LUDWIG VON Reuter, Chief of the Internment Formation, dressed with particular care on the morning of 21 June 1919. He put on full-dress uniform, hung the insignia of his highest decorations round his neck and pinned the Iron Cross First Class and his other medals to the breast of his frock coat. When he emerged on to the quarterdeck of the *Emden* he saw that Scapa Flow was also looking its best, with the sun shining from a clear sky on the morning of Midsummer's Day and a slight swell running. The air was still and warm. He was seen to pace up and down the quarterdeck 'as if waiting for something' and to be studying the interned ships through a telescope. He stopped just before 10 a.m. when his Chief of Staff, Commander Ivan Oldekop, approached him to deliver an oral report. He announced that the British battleships on guard duty had left the Flow with their attendant cruisers and destroyers. Reuter could hardly believe his good fortune. Oldekop also reported that he had seen a British newspaper of 17 June that morning which urged the Allies to have nothing to do with the German proposal to surrender all their ships in exchange for financial concessions and demanded an unconditional surrender instead. The Chief of Staff finally announced that a German destroyer, the *B98*, was due to arrive with despatches at noon the next day. Shortly after 10, Reuter ordered the international code flags 'D G' to be raised on the *Emden*. This meant that all ships were to man their bridges and keep special watch for further signals. At

10.30 he ordered another signal. Summoning the flagship's yeoman of signals, he said: 'Make: "Paragraph eleven. Confirm".' This innocuous message was the code for immediate scuttling included in his order of 17 June.

The signal was repeated by semaphore and by morse code on the signal lamps. The message took time to get round the fleet. It was sent at a time not normally used for signalling because of British restrictions, and many ships did not receive it for a while until their neighbours passed it on. The southern tail of the lines of destroyers was out of sight of the flagship and had to wait an extra hour for the order. The first acknowledgments were agonisingly slow coming in, spotted from the *Emden* just as the original signal reached the last of the destroyers at about 11.30. The response laid down in the order of 17 June was: 'Paragraph eleven is confirmed. Captain, — [name of ship].' Among the last of the Germans to know what was going on were the crew of the *Emden*. Two British drifters and a water-supply vessel were alongside the flagship, the former collecting stores, delivered by the two transports which had carried out the last reduction of the crews, for distribution round the fleet and the latter actually pumping water into the *Emden*'s tanks. Reuter dared not risk taking action on the flagship, whose crew were still unaware of his plan, while the British were alongside for fear that the excitement of a sudden burst of activity might give the game away too soon. He finally gave the go-ahead to the captain, Lieutenant-Commander Ehlers, at 12.10. There was still remarkably little to be seen one hundred minutes after the order to scuttle had gone out. Reuter became aware of the loud and deliberate tolling of a large brass bell. It came from the *Friedrich der Grosse*, anchored just two berths away from the *Emden* to the port side, and was the signal to the crew to abandon ship. At 12.16 p.m., with remarkable speed, she turned turtle and sank, the first to go down. Many German officers including Reuter noted the fact with satisfaction. 'She owed us that much for all the trouble she caused,' one of them wrote in a letter home from a PoW camp soon afterwards. Many of the destroyers, dashing to the last, hoisted the red flag, not of revolution but of the letter 'Z' in the international code, which signalled: 'Advance on the enemy.'

As Reuter anxiously paced his quarterdeck waiting for the psychological moment, the 400 boys and girls from the Stromness Higher Grade School (now the Stromness Academy) were waiting for their day's outing on board the *Flying Kestrel* (Captain Davies), the large tug-cum-tender on contract to the Admiralty from the White Star Line of Liverpool as a water-supply and general duties vessel for the internment. After the teachers had counted heads for the umpteenth time the children boarded the boat and she was under way at about 9.30 a.m. She crossed the channel between Stromness and Hoy and headed south, passing between the lines of capital ships and then along the lines of destroyers in the channel between Hoy and Fara. The *Flying Kestrel* steamed at a leisurely pace to give the children a prolonged view of the warships. They saw sailors on the decks of the ships towering over them. They got as far south as Lyness and saw the handful of British ships left in Longhope Bay after the departure of the main force – a hospital ship, two depot ships, two destroyers with smoke at their funnels and a third under repair alongside a depot ship, the *Victorious*.

One of the children aboard was Katie Watt, a senior pupil of eighteen. Sixty years later over tea and biscuits in the cheerful living room of her house in Kirkwall, Miss Watt, a retired Post Office teleprinter operator, vividly recalled the day out which none of those who went was ever to forget. 'I don't think anybody noticed anything peculiar on the outward journey at all. Some of the big ships were taking on supplies from boats alongside, and we did see one or two flags going up on the destroyers.'

James Taylor was fifteen at the time and was later to become a teacher at the school where he had been a pupil. From his retirement in the gentler clime of Sussex he kindly sent me his account of the outing as he wrote it for the now defunct weekly *Teacher's World* in June 1940. Under the headline *The strangest school journey ever made*, he wrote:

> At long last we came face to face with the German Fleet, some of them huge battleships that made our own vessel look ridiculous. [The sailors thumbed their noses.] Our teacher tried anxiously to explain that perhaps we would do the same if we were prisoners of war being stared at by a crowd of gaping school-children. We ought to feel sorry for those poor men who could no more help being Germans than we could help being Orcadians. Rognvald St Clair, who was always so smart, said he only felt sorry he hadn't brought his catapult . . . Some of the Germans . . . sat playing mouth-organs . . .

Miss Watt recalled the *Flying Kestrel* picking up a signal from the *Victorious*, reporting that the ships were sinking and ordering a return to Stromness at full speed, to land the children and return to the scene. Others said a passing drifter delivered the message. At any rate the tender put about shortly after noon and headed north. It is the staggering, nightmarish contrast between the tranquil southward pleasure-cruise and the headlong dash north through a scene of total chaos as the naval *Götterdämmerung* took place all round them that made the experience unforgettable. Mr Taylor remembers a crewman shouting: 'Look! the Germans are flying their ensigns!' Most ships indeed still had one, and those which did hoisted the white flag with the black cross at their sterns, something the Orcadians had never seen before. First they reached the destroyers. 'They were at all sorts of crazy angles,' said Miss Watt. 'Some rolled on to their sides, others went down stern first or by the bow.' Then they were in sight of the capital ships.

James Taylor's account continues:

> Suddenly, without any warning and almost simultaneously, these huge vessels began to list over to port or to starboard; some heeled over and plunged headlong, their sterns lifted high out of the water and pointing skywards; others were rapidly settling down in the ocean with little more showing than their masts and funnels, while out of the vents rushed steam and oil and air with a dreadful roaring hiss, and vast clouds of white vapour rolled up from the sides of the ships. Sullen rumblings and crashing of chains increase the uproar as the great hulls slant giddily over and slide with horrible sucking and gurgling noises under the water. The proud vessels slowly disappear with a long-drawn-out sigh.
>
> On the surface all that remains is a mighty whirlpool dotted with dark objects swirling round and round, many of them drawn inwards until they too sink from sight. Now the sea is turning into one vast stain of oil which spreads gradually outwards as if the life-blood of some ocean monster mortally wounded was oozing up from the seabed. And as we watched, awestruck and silent, the sea became littered for miles round with boats and hammocks, lifebelts and chests, spars and matchwood. And among it all hundreds of men struggling for their lives . . . Suddenly the air was rent by the lusty cheering of long lines of sailors drawn up on the deck of one of the largest German ships. They were bidding farewell to a sister-ship whose decks were now under water.

The younger children were terrified and had to be taken below and comforted or distracted by teachers and older pupils. The screams and cries from the *Flying Kestrel* composed a spontaneous and bizarre keening at the graveside of the German navy.

Miss Watt, already a young woman by then, was not frightened. 'It didn't scare me at all. I was very fascinated. I never thought anything would happen to us, although the crew were a bit worried,

I think, in case we got caught in all the turbulence.' Many eye-witnesses remember the unnatural churning of the sea caused by the sinking of the ships and the great gouts of air which broke to the surface afterwards. Others recall the weird and terrible noises the ships made in their death-throes, roaring, groaning and hissing.

Mr Isaac Moar, still sub-postmaster of Hoy at the age of seventy-five, told me how he had had a grandstand view of the scuttling from the top of a cliff near the tiny post office at the north-eastern end of Hoy, where he was already working with his family in the attached general store. He had seen the German ships arrive in the Firth of Forth seven months to the day before they sank; he was then aboard a drifter as a civilian, supplying canteen services to the ships of the Royal Navy too small to have their own. He was thus one of the very few Britons to see the beginning and the end of the internment story.

> Somebody came into the shop and said there was something fishy happening in the Flow. I walked down the road about three miles to a high point with a good view. By the time I got there, the ships were in all stages of sinking at the same time. It was just after lunchtime. The Germans were launching boats and there were drifters and cutters all over the place, rushing about everywhere. I saw the *Flying Kestrel* come racing back to unload the school-children and then it rushed back again. Then our big ships and destroyers came charging back into the Flow at full speed.

As the *Flying Kestrel* ran for Stromness, she passed the *Baden*, still at her anchorage point, which was the most north-easterly of the fleet. The last impression on the crowded minds of the children before they got ashore was the oddest, perhaps, of them all. They looked back and saw a solitary sailor in white summer uniform dancing a hornpipe on a gun-turret. The *Baden* for some unexplained reason – she was quite close to the flagship – was the last of the capital ships to start the work of scuttling herself. Thus the *Flying Kestrel*, when she came surging back to the scene, was able to take her in tow and beach her before she had time to sink. She was the only capital ship which did not go down.

The force guarding the interned ships at the time they were scuttled was the First Battle Squadron, commanded by Vice-Admiral Sir Sydney Fremantle, a conscientious if uninspired officer. His command consisted of five battleships of the 'R' class, *Revenge* (flagship), *Ramillies*, *Resolution*, *Royal Oak* and *Royal Sovereign*, two light cruisers as flotilla leaders and nine destroyers. The squadron was part of the Atlantic Fleet under Admiral Madden, who reported to the Admiralty on 22 June: 'The ships of the First Battle Squadron, with my approval, were carrying out torpedo practices on Saturday 21 June 1919, when the German vessels began to sink; the Vice-Admiral commanding had informed the Admiralty by telegraph, timed 23.50 on Friday 20 June, that HMS *Revenge* would be disconnected from the land wire between 0930 and 1800 on Saturday 21 June.' It had been quite normal throughout the internment period for the squadron based at Scapa to engage in frequent exercises, and Fremantle had planned three successive days of manoeuvres up to and including 23 June. A small guard-force – the two destroyers, *Vespa* and *Vega* – was as always left behind to keep an eye on the Germans. Two depot ships, the *Sandhurst* and the *Victorious*, and a third destroyer which was temporarily unserviceable were also in the Flow, at Lyness, where the Royal Navy's shore facilities were grouped. The usual drifters, tenders and trawlers were also on hand doing their routine errands and patrols with armed parties aboard.

On Fremantle's programme for the 21st was a simulated torpedo attack by the destroyers on the battleships. The log of the *Revenge*, preserved at the Public Record Office, contains this laconic entry for 21 June, 1300: 'German ships reported to be sinking.' A radio message from the guard-ships had broken the news. By that time just two capital ships had sunk – *Friedrich der Grosse* at 12.16 and *König Albert* at 12.54, both battleships. Fremantle ordered all his ships to return to Scapa Flow at full speed. The first were back at 2 p.m. and the last at about 4 p.m., by which time only three capital ships, three light cruisers and a handful of destroyers out of a total of seventy-four interned ships were still showing freeboard.

Meanwhile pandemonium with a strong dash of panic had broken out among the British guard-boats in the Flow when it dawned on them what was happening; the confusion was more than redoubled when the main force came back. There was indiscriminate shooting from small arms, machine-guns and the occasional shell from heavier guns aboard the destroyers. A whole kaleidoscope of incidents is recorded in the German Archives, many of them uncorroborated or distorted by memory or time. What follows is a small selection of the more serious incidents, all of them corroborated.

It is an inescapable fact that nine German internees died of gunshot wounds received on 21 June. The gravestones of eight of them, bearing that date, are still to be seen at Lyness. Sixteen more were wounded by gunfire. There is ample evidence to rebut Marder's ingenuous observation that the casualties were 'all incurred when the British fired at German ships to prevent men still on board from lowering their boats, the idea being to encourage them to undo their work'.

Miss Watt, a British witness, actually saw one man die, probably Signaller Hans Hesse of the *Bayern*. 'A drifter was towing two or three lifeboats full of German sailors. One of them got up with a knife in his hand and tried to cut his boat free of the towrope. A Royal Marine raised his rifle and shot him.' Five of the dead were killed by gunfire aimed at the boats in which the Germans abandoned ship.

As the two British destroyers, the *Vega* and the *Vespa*, alerted by emergency signals from patrol drifters, steamed into the channel between Hoy and Fara, the first German ships they reached were the tail-end of the torpedoboats which had been the last to get the order to scuttle because they were out of sight of the *Emden*. These were the ships of the Sixth Flotilla, anchored in an east–west line south of Lyness and separated from the rest by almost the entire length of the channel. This flotilla accounted for exactly half the casualties, four dead and eight wounded, incurred during the scuttling (the ninth death occurred after the crews had been rounded up and distributed among the British battleships: a German sailor failed to respond to a British order and was shot). The crews, having been the last to scuttle, were the last to abandon ship. The two British destroyers and four drifters closed in on the flotilla, opening fire with small arms. Lieutenant Zaeschmar of the *V126* was in a lifeboat with thirteen men. Three of them were killed and four wounded. The rest were ordered back aboard and forced by threats of further shooting to turn off most of the valves that had been opened (the connecting rods which enabled this to be done from deck-level had not been broken). A stoker in the lifeboat of the *V127* was shot in the stomach and died soon afterwards. The British managed to beach three ships of the flotilla, including the *V126* and *V127*. Five others sank in shallow water and were not fully submerged. Circumstances conspired to ensure that the flotilla's contribution to the scuttling was the least effective of the interned fleet. But Zaeschmar could still write from prison-camp: 'It was a sublime and yet so deeply sad feeling to see virtually nothing left of our beautiful fleet.'

A drifter put an armed boarding party aboard the battleship *Markgraf*. The captain, Lt-Cdr Walther Schumann, was helping to complete the last acts of destruction to ensure the ship sank, but he emerged waving a white flag. He was shot through the head. So was a chief petty officer. But Fleet Engineer Faustmann, senior engineer-officer of the Interned Formation, managed to stay below and complete the work, and the *Markgraf* eventually sank at

4.45 p.m., the penultimate to go to the bottom (the very last, at 5 p.m., being the battlecruiser *Hindenburg*, whose captain, Commander Heyden, coolly arranged matters so that the ship sank gently on an even keel to make it easier for the crew to escape; its masts and twin funnels stood clear of the water to mark the grave of the German fleet for eleven years).

While they were prisoners of war afterwards, Reuter and some of his officers and men set out to collect evidence that the British were guilty of 'atrocities' by virtue of the casualties they caused. This is to overstate the case. The numbers involved are remarkably small (it is even more remarkable that not a single man drowned, and that must largely be due to prompt British action) – so small that they argue an absence of system. Reuter says in his own book that the very multiplicity of targets minimised the effect of the shooting, which suggests that most of it was prompted by feelings of panic, anger and impotence. But the fact remains that a lot of ammunition was expended and several lifeboats, some flying white flags, were fired on in the confusion. Some of the shooting came from ashore as soldiers and Marines repelled boats seeking to land. On the island of Cava, a group of women armed with pitchforks and other menacing implements scared off a boatload of Germans who foolishly approached a beach.

When Fremantle returned to the Flow some of the destroyers fired salvoes, but apparently hit nothing. Their leader, Captain MacLean, is said by a number of German sources to have threatened any German commander whose ship sank with summary execution. A number of officers were lined up on the *S132* of the Sixth Flotilla and a Royal Marine firing squad was drawn up. Nothing happened. A man described as 'an English civilian' (probably an officer who had rushed back from time off ashore) boarded another destroyer, put a pistol to the head of Lieutenant Lampe, who had become separated from his men and was calling to them, pulled the trigger – and missed. The muzzle slipped and the lieutenant escaped with a nasty gash, a burn and ringing in his ears which must have sounded like sweet music for a while.

Aboard the *Emden*, the order to scuttle the ship raised the morale of the crew to a pitch it had not attained since long before the general naval mutiny. Astonishment gave way to delight as the men abandoned lunch like the crew of the *Mary Celeste* and rushed to pack a few essential items into kitbags while those best suited to the work set about sabotaging the ship. Once the *Friedrich der Grosse* went down and other ships showed signs of settling in the water, the crews of the drifters still alongside the *Emden* made to cast off. The ropes were held by German sailors until their comrades had time to toss their packs aboard and abandon ship by scrambling down into the drifters. Reuter and his staff clambered into the second drifter alongside, the one detailed for his own use on tours of inspection. Reuter decided to go ashore and persuaded the drifter to land him at a point he does not identify. He wanted to find the senior British officer ashore to persuade him to stop the shooting, which suggests the rocky bay in which he landed may have been on the coast of Hoy. It turned out that the man he wanted was aboard the *Victorious*, which had left Lyness for the scene of the scuttling. A car drew up at the landing point and the coxswain of the drifter indicated that the 'young gentleman in tennis clothes' who got out was the man to talk to. Reuter thought he seemed too young to be the shore commander; in fact, he was the commanding officer of the Orkney naval air station, based on Fara. The officer was very angry, paid scant attention to the Germans, ran back to his car, reappeared with a camera, boarded a launch and set off for the sinking fleet. The shooting continued.

The drifter took Reuter back towards the *Emden* but ran aground on a shoal somewhere in a bay on the coast of Cava. He wrote: 'The hillocks round the bay hid our ships. Only my Admiral's flag, hoisted on the *Emden* [which had no war flag left] showed above the chain of hills – it would not, would not disappear!' The surging tide refloated the drifter after an hour. As she re-emerged into open water, the first ship that came into view was the battleship *Grosser Kurfürst*, on the verge of death. She 'raised herself steeply into the air. Both cables parted with a loud snap, she fell heavily to port and capsized. The red coating of her bottom shone far across the blue sea.' It was 1.30 p.m.

About half an hour later the British destroyers returned from their exercise, throwing up huge bow-waves as they came. One of them made for the *Emden*, cut her cable and began to tow her ashore for beaching, although she was clearly sinking. The attempt however was successful and Reuter always had a pang of regret that his flagship did not go to the bottom. He asked to be taken to the *Bayern* instead. The drifter made for the battleship, pausing to take men aboard from liferafts. The *Bayern*, ensign flying, heeled over, turned turtle and went to the bottom at 2.30 p.m. Looking round, Reuter saw that a British destroyer had taken the light cruiser *Bremse* in tow and was trying to beach her. As he watched, she capsized and sank at 2.30 p.m. and the British had to cut the cable. Apart from the *Emden*, two light cruisers failed to sink. The *Frankfurt* was beached as she became awash and the *Nürnberg* was also run aground after a British destroyer crew used explosive to break her anchor-chain. The remaining five light cruisers, the most elegant of all the ships at Scapa, sank between 1.05 and 3.50 p.m. Another memorable scene took place aboard the battlecruiser *Seydlitz*. The crew stood to attention on the deck and sang the national anthem as they watched their neighbour, the *Moltke*, overturn and sink at 1.10 p.m. Then they abandoned ship and watched their own vessel go down forty minutes later. Reuter gave up the idea of going aboard a German ship and asked to be taken to the British flagship, the *Revenge*, which he reached at about 4 p.m.

Gradually some semblance of order returned to the desolate scene in Scapa Flow, the surface littered with debris, large patches of oil and abandoned liferafts. Great bubbles of air still belched from the sunken ships, the sun still shone, the patrol-boats still moved, often with lifeboats in tow. During the evening, the Germans were parcelled out among the five British battleships. Fremantle had issued an order that they should be treated with minimum courtesy, which was faithfully observed. Kitbags were violently searched and pillaged; watches and knives were stolen; men were beaten up; a man from the battleship *Kronprinz Wilhelm* was hit in the mouth and lost the false teeth he had preserved intact throughout internment 'although they were of German manufacture'; aboard the *Resolution* the ninth German sailor was shot and died later. Reuter's Admiral's cloak was stolen by a British sailor.

The Admiral and his personal staff were taken to a cabin on the *Revenge*, part of the flag officer's suite. As he was brought aboard, he had a short exchange with Fremantle, who accused him of treachery, which Reuter denied, saying, according to his book: 'It is the wish of the All-Highest that disabled ships should be sunk' and that he had genuinely believed that hostilities were about to be resumed. He was 'amazed' to learn that the Armistice had been extended by two days. Fremantle always maintained afterwards that he had informed Reuter 'unofficially' of the extension before the scuttling; the German Admiral consistently denied having any knowledge of it before the evening of 21 June. Circumstantial evidence, such as it is, supports Reuter's version, as Marder also agrees. British policy after all had always been to keep the Germans in the dark throughout internment. Short of an illegal pre-emptive seizure of the interned ships, it is also clear that the British could have done nothing. Reuter told Fremantle that he had acted on four-day-old reports from the British press indicating that Germany was resisting the peace terms. The argument petered out, filtered as it had been through an interpreter. Reuter played a game of piquet with his flag lieutenant, Schilling, and went to bed. 'I confess,' he wrote in his book, 'it was a long time since I spent such a restful night as this.'

After dark the British battleships set off for the Cromarty Firth with 1,774 officers and men from the German fleet aboard, now officially classified by Fremantle as prisoners of war. The Germans were forced to sleep on iron decks, in gun-turrents or down below without blankets, many of them still wet from immersion after abandoning their ships. Harassment went on all night on some of the British ships. Their reception ashore at Nigg, where they passed into the custody of the military, represented by the Seaforth Highlanders, had already been arranged as part of Fremantle's dispositions for seizing the German ships on the conclusion of peace. They arrived at noon.

When the squadron was at anchor, Reuter and his staff and all other German officers aboard were summoned and escorted to the deck of the *Revenge* for a unique leavetaking. Inside a square of Royal Marines and British ratings, all armed with rifles, the German and British officers faced each other, standing stiffly to attention as Fremantle appeared with a sheet of paper in his hand.

He read out the following address:

> Admiral von Reuter, I cannot permit you and your officers to leave naval custody without expressing to you my sense of the manner in which you have violated common honour and the honourable traditions of seamen of all nations.
>
> With an Armistice in full operation you recommenced hostilities without notice by hoisting the German flag in the interned ships and proceeding to sink and destroy them.
>
> You have informed my interpreter that you considered the Armistice had terminated.
>
> You had no justification whatever for that assumption.
>
> You would have been informed by me of the termination of the Armistice and whether the representatives of your nation had or had not signed the Treaty of Peace.

Indeed letters in readiness to send to you to that effect as soon as I had received official intimation from my Government were written and signed.

Further, can you possibly suppose that my squadron would have been out of harbour at the moment of the termination of the Armistice?

By your conduct you have added one more to the breaches of faith and honour of which Germany has been guilty in this war.

Begun with a breach of military honour in the invasion of Belgium it bids fair to terminate with a breach of naval honour.

You have proved to the few who doubted it that the word of the New Germany is no more to be trusted than that of the old.

What opinion your country will form of your action I do not know.

I can only express what I believe to be the opinion of the British navy, and indeed of all seamen except those of your nation.

I now transfer you to the custody of the British military authorities as prisoners of war guilty of a flagrant violation of Armistice.

An interpreter translated this to Reuter as the Germans stood with expressionless faces, their eyes focused on the middle distance as parade etiquette required. The German Admiral's reply, delivered in his own language, was short. He said to the British interpreter:

Tell your Admiral that I am unable to agree with the burden of his speech and that our understanding of the matter differs. I alone carry the responsibility. I am convinced that any English naval officer placed as I was would have acted in the same way.

The ceremony, which Reuter suspects was got up for the benefit of the reporter representing *The Times* who had managed to witness it, apparently brought lumps to the throats of Germans and British alike. The Germans clicked their heels and were led away to captivity.

The men who had been shot the previous day were the last casualties, and the survivors the last prisoners captured, in the First World War, which officially ended with the signing of the Treaty of Versailles on 28 June.

In his account, Reuter, a romantic if ever there was one, records this reflection on what happened on Midsummer's Day 1919:

> Now the German fleet lies on the cold seabed, and in my thoughts I shake the hands of the fine comrades who brought off this last patriotic deed of the navy.

Fremantle's remarks, coming as they did from the man under orders to seize the interned ships on the termination of the Armistice (whether by the conclusion of a treaty or the resumption of hostilities), doubtless relieved his embarrassment. They seem distinctly ironic now, delivered as they were on the deck of a ship bearing the name HMS *Revenge*. Like any other British naval officer, he must have known the story of Sir Richard Grenville and that earlier ship of the line which had bequeathed her name to his flagship, immortalised by Alfred Lord Tennyson:–

> Sink me the ship, Master Gunner – sink her, split
> her in twain!
> Fall into the hands of God, not into the hands
> of Spain!

CHAPTER TEN

Aftermath

REACTION TO THE greatest act of material self-destruction in the history of warfare was decidedly mixed. The British were publicly indignant and privately relieved; the Germans officially regretful while protesting their innocence, but unofficially proud. The French were furious and vengeful, while the Americans shrugged their shoulders. We have already seen how the Allies had for months been unable to agree on the ultimate fate of the interned ships. For better or worse, Reuter had disposed of a bone of contention once and for all by his spectacular act of sabotage. The Royal Navy's reaction is perfectly summarised by a sentence in a letter written on 22 June from Admiral Wemyss, of the British delegation at Versailles, to a fellow Admiral then in Paris: 'I look upon the sinking of the German Fleet as a real blessing.' The British Prime Minister, Lloyd George, no doubt privately felt the same.

For all their carefully concealed satisfaction that a large number of modern warships, which could have upset the postwar balance of power at sea, had been destroyed, the British simulated anger quite convincingly, and secretly ordered diplomats and Secret Service agents to leave no stone unturned in the search for proof that Reuter had acted on direct orders from Berlin. The Germans got wind of this search in Copenhagen. They learned from their own diplomatic sources that British agents in the Danish capital had been ordered to find out whether Reuter was in touch with Berlin over the scuttling. The information was to be obtained by any means available and regardless of cost. Official proof was to be sent or, failing that, information on where it was kept or could be obtained. Enquiries were to be made in the German naval bases of Kiel and Wilhelmshaven.

The highly uncharacteristic British readiness to write a blank cheque for a piece of information contrasts so strongly with the normal financial circumspection, not to say stinginess, of the British ruling class that it can indicate only how important proof of collusion was regarded at the time. Their complete failure to find it then and the absence of proof even long after the records on both sides became available to the public acquires special significance by the same token. The Admiralty started an internal investigation on the very day of the scuttling and sent a telegram to Admiral Madden as C-in-C of the Atlantic Fleet that evening demanding urgently a full report on the sinking and a copy of the orders for guarding the German ships. A commission of enquiry was set up and took evidence from British and German sources (though not from Reuter himself). A report dated 23 June 1919 by the Secretary of the Allied Naval Council made no bones about saying, 'I told you so.' The opening paragraph includes the words: '. . . it will be seen that the British and French Naval Delegates never wavered from their opinion that the German surface warships should be surrendered and not interned, and only accepted the internment finally as a decision of the Heads of their Governments and after reiterated protests'.

In the end, after nearly six months of fruitless enquiries, the best the British could do by way of producing evidence of collusion was to quote the letter of Admiral von Trotha to Reuter of 9 May 1919 out of context. The letter was found at the beginning of December among documents in the ship's safe aboard the unsunk *Emden*. No less a personage than the Director of Naval Intelligence, Commodore Hugh Sinclair, took the extraordinary step for such a normally self-effacing functionary of calling a press conference, itself a rare enough manifestation in those days, on 3 December, at which he produced this mildewed rabbit from an otherwise manifestly empty magician's hat. He pointed out that two steamers had arrived in Scapa Flow from Germany on 17 June, the day Reuter drew up the order to scuttle, a copy of which had been found on the *Emden*. Sinclair speculated that the order from Berlin to do this had been brought by the steamers, perhaps concealed in a tin or a loaf! He also announced that Reuter would remain in captivity as 'one of those required in connection with atrocities'. It was at best a flimsy case, but the Allies felt they needed it to justify their claim for reparations for the scuttled ships. In a Note of 8 December, they accused the German government outright of having specifically ordered Reuter to sink the interned fleet, quoting the Trotha letter out of context. Five weeks earlier they had demanded as compensation the surrender of five more light cruisers and 400,000 tons of dockyard equipment, including dry docks, cranes, tugs and dredgers and made it clear that the crews of the sunken ships would be kept in PoW camps as hostages for the fulfilment of these additional peace terms.

A neutral verdict on the significance of the Trotha letter is to be found in the serious and influential newspaper *Nieuwe Rotterdamsche Courant* of the Netherlands, which took no part in the war. The paper commented at the time of the revelation of the letter that nobody without prejudice could possibly read it as an order or even a hint to scuttle. This is therefore perhaps as good a time as any, before turning to other aspects of the reaction to the scuttling, to dispose of the question of whether Reuter jumped or was pushed.

The Admiral himself expected to be put on trial for what he had done and spent much of his time in imprisonment preparing his case. He thought the British might try him for war-crimes, breaching the Armistice or even the quaintly named offence of barratry (fraud or deceit by a ship's master and/or crew against the owner); or else the Germans might court-martial him for destroying state property without due cause. The Allies considered arraigning him but dropped the idea by 21 July; even the discovery of the Trotha letter did not serve to revive it, though he was still in captivity in England. Reuter's repeated public claim of sole responsibility for the sinking was made with a possible trial in mind. *The Times* reported within days of the scuttle that Reuter was to be charged with a breach of the Armistice. Reuter says in his book: 'Accepting the idea that I would immediately be arraigned before an English judge, I gave no more information on my motives. I did not want to be committed to anything in writing.' The fact that the British took no such action, even when the French and the Italians were openly blaming the British Admiralty for the scuttling and American agents were claiming to have heard of British connivance in it, is eloquent enough in itself. It could of course be argued that the British decision not to prosecute reflected official embarrassment over the Royal Navy's failure to prevent the scuttling, or fear of imposing unnecessary strain on relations with the wartime Allies, rather than fear of being unable to make a case. But when the decision is viewed in conjunction with the protracted search for evidence, the demonstrable failure to find it and the ensuing distortion of the Trotha letter, it seems clear that it was based on an unspoken recognition that there was no case to answer.

Such arguments do not of themselves settle the question. It has been shown that a scuttle was uppermost in the minds of the German Admiral and his officers and colleagues at home from the moment the Armistice terms became known. Then there was the standing order in the Imperial Navy that no ship was ever to be handed to an enemy, a precept which any naval officer of any nationality with any sense of honour would take for granted and one which Reuter often quoted as a principal justification for what he did, given his claim that he believed hostilities were about to be resumed. It is entirely fair and reasonable to assume that the subject of an eventual scuttle was discussed at length during Reuter's long leave in Germany in December 1918 and January 1919; it is well known that the interned fleet and the German authorities were able to communicate with each other without the knowledge of the watching British by such channels as officers going home permanently or on leave or through the commanders of German warships and steamers delivering supplies or mail to Scapa Flow (Trotha's letter is a prime example). The German penchant for putting everything down on paper notwithstanding, there was actually no need to conceal documents in hollowed-out loaves. There was always the possibility of entrusting a chosen officer with a oral communication. This is precisely what Grand Admiral Erich Raeder, who was to be head of the German navy from 1928 to 1943, records in his memoirs as having taken place. He was on the staff in the German Admiralty at the time and there is no reason to think he had any reason to distort or suppress the truth in what is little more than an incidental reference to the scuttle. He says that Trotha impressed it upon Oldekop (Reuter's Chief of Staff) orally during a visit by the latter to Germany on leave that the ships must be sunk regardless. Raeder also says that the same message was sent to Reuter via the captain, a Commander Quaet-Faslem, of one of the warships sent from Germany to Scapa Flow with despatches. This officer apparently also delivered the message orally, but, infuriatingly, Raeder gives no further details. There is thus no evidence here that Berlin ordered Reuter to scuttle at a specific moment in a specific response to the presentation by the Allies of the peace terms at the beginning of May. An even sloppier piece of 'reporting' is to be found in the autobiography of Lieutenant-Commander J. M. Kenworthy (later Lord Strabolgi). He baldly states that when it became known that France and Italy wanted a share of the interned ships and the Americans approved, because this would reduce British naval supremacy and with it British power, it was common knowledge that the Royal Navy hinted openly to the Germans that a 'gallant hara-kiri' would not be received without sympathy. While this may reflect private British attitudes in some of the highest places, there is not a shred of evidence to support it.

Finally, therefore, we are left with the British failure to prove collusion, the consistent British policy of cutting off the interned fleet from the outside world as far as possible by stringent censorship, the German standing order against surrendering ships to the enemy – and Reuter's character as an officer of the old school with a clear concept of military honour, made all the keener by his consciousness of the disgrace which the German navy brought upon itself in the mutiny that immediately preceded internment, which can only have made a man like him all the more anxious to salvage the reputation of the fleet by sinking its helpless and rusting ships under the eyes of the enemy. It is entirely possible to assume that Reuter closed his mind to the innate absurdity of his own belief (or subsequent rationalisation) that hostilities would be resumed after such a long armistice. One can go further and conclude that he would have sunk the ships in almost any circumstances rather than hand them over. But what the evidence does not allow one to conclude is that he acted in response to a specific order. Not only is Reuter entitled to the benefit of the very strong doubt which exists on the point; it is also absolutely clear that such an order was unnecessary for such a man in such a position. The whole argument thus becomes irrelevant except in the context of the Allied self-justification in demanding reparations for the lost ships, an imposition which in the end was supported only by the most powerful argument of them all: might is right and winner takes all.

It is difficult not to feel sorry for *Kapitänleutnant* Hengstenberg, commander of the destroyer *B98*, which left Wilhelmshaven on 21 June with mail for the internees. He reached the rendezvous out in the North Sea, was met by the destroyer HMS *Westcott* and conducted to Scapa Flow by a roundabout route so that he saw as little as possible of the main anchorage. But he spotted some beached torpedoboats, probably of the Sixth Flotilla, and it began to dawn on him what had happened. 'It looked as though a scuttle of the German ships had taken place,' he reported. On his arrival on 22 June, his ship was seized by an armed boarding party. Seeing this as a clear breach of the Armistice, Hengstenberg immediately issued an order to scuttle, but the British were able to prevent it. The German sailors were declared to be prisoners of war and their ship was thoroughly looted. A week later, the men were sent home on a German steamer, but the British retained the *B98* by way of reparation, claiming that there was 'a state of war in Scapa Flow' because of the scuttling. In demanding the repatriation of the crew the Germans argued: 'The destruction of one's own weapons could not possibly be regarded as hostility . . .'

The rough treatment handed out to the crews of the scuttled ships in the immediate aftermath of the sinking by the Royal Navy has already been described. The unofficial hostility continued after they were put ashore at Nigg. Most German accounts agree that the Seaforth Highlanders behaved in a generally correct way, but elements of the regiment did not, looting such possessions as some of the Germans still had left to them after the navy's depredations, sometimes at bayonet point. After spending their first night ashore in the camp at Nigg, the Germans were moved southwards by train to prisoner-of-war camps in the north of England. There were ugly scenes at some of the stations on the way, such as Invergordon and Lancaster, where hostile crowds, including off-duty soldiers, women and even children, booed, spat, threw punches and stones, and blocked the line. When he reached Oswestry in Shropshire, Reuter was taken under escort to a bank to change money. A crowd of about 1,000 people swiftly gathered to shout abuse, a woman struck him on the shoulder and a man threw a lump of coal in his face.

But the long, slow train journey had its compensations. After seven months cooped up in uncomfortable and dirty ships in the glowering shadow of Hoy, the leisurely trains passed through the magnificent Highland scenery in splendid weather and many aboard were reduced to tears of joy by the views, despite such discomforts as having to use their own seaboots to store water for the journey.

While the British were hostile, the German soldiers already imprisoned at the camps cheered the sailors when they were marched in. Although the British press did not approve, some newspapers went so far as to acknowledge that the Germans might have had the right to sink their own ships or at least that the act of destruction was not beyond comprehension. The German press naturally took a different view, recording the scuttling under such headlines as: 'The last heroic act of the German navy' and '1½ milliards (thousand million) at bottom of sea'.

Among the general public in Germany the scuttle caused a brief flicker of interest and approval in contrast with the general unpopularity of what was still very much the junior service. The support aroused by the powerful and enthusiastically supported Navy League before the war had been eroded by the traditional bias towards the army, which delivered a stream of victories however pyrrhic, as well as by the contrasting failure of the navy to provide its share of glory, and by the mutiny at the end and by the disruptive social influence of the navy's presence along the German coasts. Many mothers would not allow their sons to wear the traditional sailor-suits of the time. The Kaiser was not displeased with the final act of the service he had so uselessly built up. He contacted the Navy Office from his exile in the Dutch castle of Doorn on 26 June asking for details of casualties and survivors of the scuttle.

The Left in Germany was hostile and suspicious, as a commentary in the radical magazine *Die Weltbühne* of Berlin dated 3 July 1919 illustrates. The author was the left-wing commentator Heinrich Stroebel.

> The deed of Scapa Flow could become of most evil portent.
> It was the extension of militarism against the civil power of

> the Republic. The German press has called this military sabotage of peace a 'heroic deed of German sailors'. Should there not be deeds more heroic than the sinking of dead ship-material? Be that as it may, once armistice and peace are concluded, so soldierly heroism, which tears up treaties as scraps of paper, becomes a crime against the state. The naval crews may not have known that; the commanding Admiral must have been aware that his act was insurrection against the German Government. And Rear-Admiral von Reuter even said himself that he regarded the scuttling of the ships as the execution of an order of the Kaiser!

Clemenceau, the French Prime Minister, protested on behalf of the Allies about the scuttling (and an outbreak of burning French flags in Berlin) in a Note to Germany of 25 June 1919, delivered to her delegation at Versailles. The German government replied with a flat denial of collusion: '. . . the scuttling . . . was undertaken without the knowledge or wish of any official or military authority in Germany whatsoever'. A German protest over the looting and shooting by British sailors after the scuttling was answered by a British denial of the former and an admission of the latter, which was put down to individual sailors not hearing orders. A British naval commission of enquiry interviewed officers and men from the interned fleet, but by the middle of July 1919 Reuter, who was never questioned, had concluded that he would not be put on trial, either by the British or by the Germans. The imprisoned Admiral felt it incumbent upon him to make a report on the scuttling to his superiors and cast about for a way of doing so without the British being able to intercept it. In the end he fastened upon a Lieutenant Lobsien of the German Army Reserve, who was about to be repatriated to vote in the plebiscite on the future of Schleswig-Holstein. For eight days Reuter and Lobsien went over the report, dated 15 July, until the lieutenant had it off by heart for oral delivery to the

German Admiralty. He was able to pass it on only in the middle of August because his repatriation was delayed. Reuter explained:

> The reasons for my action at Scape Flow on 21 June 1919 were the two instructions, (1) a naval commander-in-chief in foreign waters who is not in touch with home is to act on his own initiative as best meets the requirements of the state and the honour of the navy; (2) German ships must on no account be allowed to fall into the hands of the enemy in time of war. [The ships would be sunk in the event of an attempt to seize them or if Germany rejected the peace terms; they would be surrendered if Germany signed the treaty.]
>
> On Friday, 20 June, during the afternoon, I learned from *The Times* of 16 June the official text of the German counter-proposals. In these, the German fleet was to be used as an object of barter. That evening I had a wireless message transmitted to the German Government asking that the officers should be relieved of their duty before the surrender.

In fact the message could only be passed to Fremantle on the morning of the 21st and was never sent. After that Reuter saw *The Times* of 17 June, containing the official text of the final response of the Allies to Germany, including a refusal to accept any of the German counter-proposals, not least that the interned fleet should be a bargaining counter. The Germans were given five days (from 16 June) to accept.

> I had to conclude that if the Peace Treaty were not signed by Saturday 21 June, a state of war would once more apply.
>
> Judging by the tone of rejection of the final Note of the Entente, and by the speech of the Government spokesman in the German National Assembly and the attitude of the German press, I was entitled to assume that the Treaty would

> not be signed, and therefore that a state of war would be resumed on the Saturday. In this context I should like to say that I received no information of any kind, either from the German or the English Governments, about the extension of the Armistice period as given in the Note of 16 June, or of any views the German Government held in general terms. My only sources of information were the English newspapers, which I received as a rule four days after their publication. In the belief that the state of war had been resumed, I gave the order to sink the interned German fleet on Saturday 21 June. I first learned of the extension of the Armistice period to Monday 23 June aboard the English flagship *Revenge* during the afternoon of 21 June, from the interpreter to Vice-Admiral Sir S. R. Fremantle. In my opinion, therefore, no reflection can be cast on me for a breach of the Armistice.

Cut off from all sources of immediate information, Reuter was obliged to carry out standing orders, he said. It was not a case of 'using my own initiative' as Clemenceau claimed in his Note, but only of executing an order 'which had not been cancelled'. The departure of the British guard-force had been coincidental. What was more, Article XXXI of the Armistice had said only that the ships were not to be damaged *before* delivery: 'what must not be done with the ships after delivery is not stated [a splendid example of German legalism, this!] . . .' As a result of this omission, a loop-hole remains which of itself justifies the scuttling, especially as it was a case of German property and not the enemy's.

Reuter was puzzled by the fact that he was not put on trial in the light of the Allied demand for reparations for the ships he had sunk. Even if the British were not interested in a trial for fear of a reopening of the question why the Royal Navy had failed to prevent the scuttling, the Admiral thought the German government might press for his trial before an international, a German or even a British tribunal in the hope of getting a clear verdict on the scuttle. There was always the possibility that a court might be driven to conclude that Germany was not responsible for Reuter's decision, which would make it harder for the Allies to press for reparations. 'In any case,' Reuter wrote in his book, 'I reject most emphatically the blame assigned to me for the subsequent surrender of the German harbour material.'

Reaction among ordinary people in Orkney to what the Germans had done was entirely favourable, according to postmaster Isaac Moar. 'As far as Orkney was concerned, it was a good thing. We were all glad he did what he did. The Allies were arguing among themselves about what to do with the ships. He did a good service to peace – and provided a lot of local employment later on!' Katie Watt said: 'Everybody thought it was the best thing that could have happened. It stopped an awful lot of squabbling about what to do with all those ships.'

Reuter was the most senior German prisoner of war in Britain, and he occupied his time with efforts to obtain repatriation for his men. The Treaty of Versailles provided for the repatriation of all prisoners upon ratification. Controversy in Britain and in its press about the high cost of keeping German prisoners at a time of shortages left over from the war ensured that thousands of German soldiers were sent home as soon as was practicable. But the men from the scuttled fleet were kept behind. In autumn 1919 the Allies presented their demand for reparations for the lost ships in detail, in the form of a Protocol to the Treaty. Five light cruisers were to be surrendered within 60 days; 400,000 tons of dock equipment within 90 days (a list of such items to be ready in ten days); the unfortunate *B98*, which arrived in Scapa the day after the scuttle, was to be retained; and the crews of the lost fleet were to be repatriated on the fulfilment of the first two demands at the latest. In other words the crews effectively became hostages for German compliance with the protocol. This led Reuter to mount a campaign for their release. The Admiral was angered by the fact that the German press seemed indifferent to their fate and that the government was equally inactive. He drew up a telegram pressing the government to take up the matter, and sent copies of it to the German press in advance to ensure that his message was not conveniently 'lost' in Berlin. A British General who happened to be visiting Reuter's camp at Donington Hall in Derbyshire courteously and helpfully redrafted the telegram in good English so that it would pass the censors more quickly. A German army captain due for repatriation at the time agreed to organise a campaign in Germany to rouse the government by sending letters to newspapers and parliamentary deputies.

Reuter wrote a letter to the British Prime Minister, Lloyd George, pressing for repatriation of the crews, on 13 October. It was delivered through the good offices of the Swiss Legation in London, the Swiss being the protecting power for German prisoners. A copy is in the German Archives, three pages of poor-quality lined foolscap paper covered in small, neat, stiff but highly legible hand-writing in unfaded blue ink; Reuter wrote the courtesy copy himself for Admiral von Trotha as head of the Admiralty in Berlin.

In the letter, Reuter once again sets out his argument about having acted in good faith on the basis of out-of-date information. He had been kept in the dark by the German government and the Royal Navy guard-force. According to what he had read, 'the German Chancellor had even declared in the national assembly that he would rather his hand wither than sign the Peace Treaty . . . as an officer, I too was unable to think it possible that such a Treaty could be signed.' (Among the things Reuter did not learn until later about the hectic days preceding the signature of the Treaty on 28 June was the change of Chancellor, which made it possible.) Reuter told Lloyd George that, given his reasonable assumption that war was about to break out again, he was governed by two service regulations: one required a commander in foreign waters to act independently according to his best understanding of the interests of the state and the honour of the navy in case of emergency; the other said: 'Ships put out of action must not fall into enemy hands.' He added: 'The ships placed under my orders were disabled; the only thing left for me to do was to sink them. This I did . . . I am convinced that English commanders have the same instructions and would do as I did.' There had been no wilful or guilty infringement of the Armistice. He had acted as 'an officer and man of honour'. The exclusion of the Scapa crews from repatriation was discriminatory: 'We, who have after all only carried out our duty, will not be treated with justice, fairness and the chivalry of war, but will just be serving as victims of a kind of vengeful feeling.'

Reuter had in fact started his letter with a false assumption. 'I have just learned that I, Lieutenant-Commander Wernig and Lieutenant Schilling [Reuter's flag-lieutenant] and six ratings of the German fleet recently interned at Scapa Flow, have been excluded from the general scheme of repatriation . . . from Donington Hall. I conclude from this fact that a similar procedure will be adopted with the remaining officers and men of the interned fleet.' In fact the cause was a transport strike which held up the repatriation of many of the prisoners, but all the officers at least of the scuttled fleet were repatriated by the end of October, with the exception of Reuter and the two he named in his letter. Lloyd George was not opposed to completion of the repatriation but the French were less keen to hurry. As this was the time the reparations issue came to a head, Reuter thought he would be put on trial in support of the Allied demands and, as we have seen, could not understand why he was not. On 9 December 1919, the continued imprisonment of the Donington nine was raised in the German National Assembly; two days earlier came the revelation that the text of the order to scuttle and von Trotha's letter to Reuter had been found aboard the *Emden*. On 14 December, the German government formally requested the release of the last prisoners in a Note. On 22 December the Allies responded with a demand for early delivery of the last five modern light cruisers left to the German navy. Reuter meanwhile, on the very day of his promotion to Vice-Admiral, 12 December, had written a second letter to Lloyd George emphatically denying that in scuttling the ships he had been acting on a direct order from Germany and rejecting the deliberate misinterpretation of the Trotha letter to him as an order, or even a hint: the letter had been irrelevant to his decision.

Germany eventually signed the Protocol on reparations on 10 January 1920. In the early hours of 29 January, Reuter and the handful of prisoners left at Donington Hall were told that they were going home and were taken to a special train which carried them to Hull. The same evening the German Steamer *Lisboa* set sail for Wilhelmshaven with the last German prisoners of the war to be repatriated. The steamer reached the mouth of the Jade, where the last voyage of the German High Seas Fleet had begun 14 months earlier, at dawn. A flotilla of German destroyers greeted the ship, and on arrival in the port which had been the principal base of the fleet Reuter received an emotional welcome. A band played the musical honours appropriate to the formal reception of an Admiral; Trotha, accompanied by other officers, was on the dockside, as were veterans' groups, troops and civilians. They cheered the only man in the history of the world who had scuttled a navy.

After coming home, Reuter collaborated in the preparation of an official German government report on the rough treatment meted out by the British in the immediate aftermath of the scuttle. The resulting *Memorandum on British breaches of international law committed against the crews of the German fleet scuttled in Scapa Flow* was published by the naval administration of the Ministry of Defence on 24 February 1921. It dealt with the outbreaks of shooting as the German ships were sinking, the bullying and robbing of German sailors as they were detained by the Royal Navy and isolated incidents in the PoW camps. The evidence was taken from sailors, some of whom had written down their experiences on scraps of paper while in PoW camps and hidden them for later reference. These and a small mountain of typewritten statements taken from sailors are to be found in the German Archive. The report and selections of evidence appeared in the German press in March 1921. The British consistently denied brutality although they eventually admitted individual cases of looting of German sailors' property.

What became of the remarkable officer and gentleman who took that unique decision to destroy a fleet? Five months after his return he was dismissed from the service because there was no job available for a Vice-Admiral of fifty-one in Germany's residual miniature navy. A letter from Defence Minister Geissler, dated 26 June 1920, said:

> The events which have occurred and the measures thereby made necessary make it no longer possible, to my regret, to employ the services of Your Excellency in the navy in a position appropriate to your rank. You are therefore respectfully requested to hand in your resignation. I should not like to allow this occasion to pass without expressing thanks to you for the outstanding services you rendered the Fatherland in the navy. History will record your name as that of a man who contrived to make many German hearts beat faster in the dark days of 1919.

Reuter had to give up his navy house in Wilhelmshaven and moved to Schloss Gauerwitz near Dresden to write his book on the scuttling, *Scapa Flow: Grave of the German Fleet*. In 1923 the family moved again, to Potsdam, where Reuter became a councillor. The former Kaiser, the man who had caused the useless navy which Reuter destroyed to be built, did not forget the Admiral in his peaceful retirement. From his exile at Doorn in the Netherlands, by then in its 21st year, Wilhelm sent the following telegram on 9 February 1939:

> On your seventieth birthday, my dear Admiral von Reuter, I express, also in the name of Her Majesty the Empress, our warmest congratulations. Your war exploits as captain of my cruiser *Derfflinger* and as commander of the Reconnaissance Groups of the High Seas Fleet constitute a scroll of fame in the history of German warfare at sea. With your act at Scapa Flow you saved the honour of the Imperial Navy. You thereby earned the ineradicable thanks of myself, the navy and the entire German nation. – Wilhelm.

As the shadows of another war darkened, Vice-Admiral Ludwig von Reuter was promoted honorary full Admiral on the occasion of the 25th anniversary celebration of the Battle of Tannenberg, where the Germans under Hindenburg crushed the Russians, on 29 August 1939. Four years later, on the way to a meeting of Potsdam Council, Reuter collapsed and died of a heart attack. The date was 18 December 1943, and thus he was spared the grief of seeing a second German defeat in his lifetime. He was buried next to his youngest son, killed in Poland while serving as an Army Ensign.

So much for the man who sank a fleet: it is time to consider what happened (and what is still happening) to the ships.

PART IV

Picking up the Pieces – The Saga of the Salvage

They that go down to the sea in ships: and
occupy their business in great waters:
These men see the works of the Lord: and his wonders in the deep.

– Psalm 107:23, 24 (Book of Common Prayer)

CHAPTER ELEVEN

Unfinished Business

THE UNPARALLELED MARITIME hara-kiri carried out by the Germans in Scapa Flow is not the end of the story of the High Seas Fleet. It was to be followed, fittingly enough, by the greatest salvage operation in history, and this spectacular last act in the drama, still not quite finished, belongs to the British. As we have seen, the scuttle was less than complete. The battleship *Baden*, the light cruisers *Emden*, *Frankfurt* and *Nürnberg*, and a total of eighteen destroyers were beached by the Royal Navy in the chaos of 21 June 1919 before they could sink, a total of twenty-two. But ten battleships, five battlecruisers, five light cruisers and thirty-two destroyers went to the bottom, a total of fifty-two out of the seventy-four ships interned in November 1918: about 70 per cent in terms of numbers but about 95 per cent in terms of tonnage.

Two days after the scuttle, the British Admiralty said they would be left to rot: 'Where they are sunk, they will rest and rust. There can be no question of salving them.' Naval salvage experts inspected the scene the next day and unwisely concluded that the great collection of wrecks was not a danger to navigation, something that was soon disproved when small boats began to run aground on some of them. Not surprisingly after such a long war involving the use of so much material, there was no shortage of scrap metal and thus no incentive to add to the glut by gratuitously taking on the formidable and expensive task of raising sunken warships.

In 1922, however, local initiative succeeded in salvaging one of the destroyers sunk in shallow water and bringing it to Stromness for breaking up. In April 1923 the Admiralty changed its mind and indicated that it was prepared to consider tenders for raising some of the ships. In June the first contract was agreed with Mr J.W. Robertson, leader of the Zetland (Shetland) County Council, giving him the right to raise four destroyers from the shallower waters near Fara Island. Scuttled destroyers came cheap, even in terms of the money-values of the time. The Admiralty charged £250 each. Later, when the big ships were sold to other contractors, it was possible to buy a battleship for £1,000 (the last that were sold fetched the still not unreasonable price of £2,000 each).

Robertson bought two surplus concrete barges from the Admiralty which between them could lift 3,000 tons. By placing a framework of steel girders across the gap between the two barges, he had a lifting platform. His company, the Scapa Flow Salvage and Shipbreaking Co. Ltd, then patented a specially designed huge buoyancy bag, not unlike a barrage balloon, nicknamed a 'camel'. One of these was to be secured to each side of the sunken destroyer, steel belts were passed under its keel, chains were attached to either end, with the idea that the ship would be pulled out of the mud by winding on the chains, whereupon, once free of the bottom, the buoyancy bags would help to float it to the surface on an even keel. It took a long time to prepare, but the method worked.

In January 1924, before Robertson had lifted his first destroyer, Mr Ernest Cox arrived in Orkney and announced that he had bought the *Hindenburg*, the *Seydlitz* and two dozen of the smaller ships. Eventually he was to buy the rest, earning for himself the sobriquet of 'the Man who bought a Navy', the title of a book about him and his work by Gerald Bowman. His company, Cox and Danks Ltd, won the contract against competition from several other firms which wanted to exploit the underwater 'steel-mine' the submerged fleet represented now that the scrap-metal market had picked up. Cox was allowed to use the facilities, such as they were, of the naval base at Lyness, virtually abandoned soon after the end of the war. He installed wireless and cinema facilities for his workforce, which was put up in the bleak and down-at-heel naval huts at the little port. His contract was confirmed on 12 February 1924.

Cox was a classic self-made man, one of the last of the Victorian self-help school, a rough diamond and a believer in the principle that 'where there's muck, there's money'. The eleventh son of a struggling (and no wonder) Wolverhampton draper, Cox left school at thirteen and took a job as an errand-boy. In his spare time he indulged his voracious appetite for mechanical engineering books, borrowing them from the local Mechanics' Institute Library. At twenty, he was chief engineer at a Hampshire power station. Later he moved to Lancashire, married the daughter of the owner of a steelworks, became a partner in the firm and developed it. Cox was a man of blunt speech which often passed the point of rudeness, a brave and exceptionally stubborn individual with a quick temper. He was short on modesty and long on hard work. In 1913 he set up Cox and Danks with his wife's cousin, who put up the money while Cox contributed the expertise. They made a lot of money from the war, manufacturing shell-cases. After the war, Cox bought Danks out and turned to ship-breaking at a yard on the Isle of Sheppey in the Thames estuary. He bought a German floating dock, part of the reparations for the scuttled fleet of Scapa Flow, from the Admiralty for £24,000, and successfully broke up and sold off two scrapped British battleships. Short of work, he took up the suggestion of a Danish business contact to turn his attention to the wrecks in Scapa Flow. Ignoring expert scepticism about the depth of water over most of the ships, Cox eventually invested about £100,000 in the operation, buying tugs, cranes, pumps and diving gear. Cox's best move in setting up at Scapa was, following a chance encounter on holiday, to employ a diver called McKenzie as his chief salvage officer. McKenzie proved to be a technical master who extended the frontiers of marine salvage, supported by Cox's energy, determination and money. The only important element missing in the equation was business efficiency, about which

Cox knew nothing and apparently cared less. It was to prove his Achilles' heel, but not for a long time yet. Cox hired divers and skilled workers from many parts of Scotland, but for the rest the labour was local and very grateful for the work at a time of severe economic difficulties for the Orkney community. At its peak, the Cox and Danks operation employed 200 men, who lived at the Lyness camp.

When Cox started work in spring 1924, he soon found that others had preceded him to the accessible wrecks. Like other communities who live by rough seas, the locals regarded wrecks as fair game. Fishing boats put out at night and stripped the upperworks of the big ships that were partly above water of everything small enough to move. Watchmen were retained from now on.

Cox decided that his first target would be the destroyer *V70*, submerged upright in only fifty feet of water about half a mile from Lyness. He had his floating dock cut into two L-shaped halves, which were placed on either side of the wreck's position. The technique he used to bring it ashore was completely different from Robertson's and somewhat simpler. Lifting chains were passed under the hull and tautened at low tide. As the tide came in, the destroyer in its cat's cradle of chains hung between the floating docks, which moved her inshore. The process was repeated during successive tides until the boat could be beached. But errors were made before the technique succeeded. The first set of chains, used by Cox against advice, were too weak and snapped one after the other with a terrifying and explosive whiplash effect. He had to order much stronger steel cables. The boat was beached after five lifts on 4 August 1924, in six weeks of work. Finding that the ship had been stripped of much valuable scrap by illicit divers beforehand and mindful of a collapse in the scrap-steel market, Cox turned the *V70* after basic repairs into a floating workshop for use on other wrecks. Having learned from his mistakes, Cox set about lifting one destroyer after another and at an accelerating rate. On 29 August he lifted his third while the smaller, rival Robertson operation lifted its first, the *S131*.

Robertson completed the raising of his four destroyers by the end of 1924 and took no further part in the salvage operation. What was left of the High Seas Fleet now belonged exclusively to Cox's company. Work proceeded on the destroyers throughout 1925: in June the first fatality occurred when a worker was crushed by a falling crane; in September Cox bought a second and larger floating dock, which he used to adapt his lifting technique. The original half-docks now lifted the wreck into the submerged new one, which was then floated and taken ashore at Lyness. The average rate of lifting destroyers surpassed one a month overall, with twenty-four being raised in twenty months. The fastest so far, the *S65*, took four days to recover from start to finish of the salvage operation, which must have been some kind of record. Cox sold the resulting 23,000 tons of scrap for £50,000. The only important setbacks were the chain-snapping incident with the first wreck, the death of the worker and the capsizing of the new dock during an attempt to raise a large destroyer. The last of Cox's two dozen destroyers, the *G104*, broke the surface on 30 April 1926. But all that effort and success amount to a mere *hors d'oeuvre* compared with the epic struggle that was now to begin, the recovery of the capital ships.

The first to receive Cox's attention was the most visible, the battlecruiser *Hindenburg*, so considerately sunk on an even keel by her captain in 1919, and still upright on the bottom in seventy feet of water, her funnels, mast and superstructure above the surface, a rusting memorial to the lost fleet, that could be seen for miles. She was the largest (though not quite the heaviest) ship in the fleet and in the Imperial Navy. Cox and McKenzie realised that a new technique would be needed to lift this monster, at 26,000 tons exceeding the combined displacement of all twenty-four destroyers taken together. Cox ordered his second floating dock cut into two L-shaped halves like the first. The plan was to patch all the holes in the hull, including the seacocks which had been opened to sink her, pump her clear of water, and use the four half-docks to help beach her. By an early stroke of luck the ship's plans were discovered in legible condition aboard. The divers who went down into the submerged hull found an eerie world below, full of enormous marine growths and unimaginable shapes just visible in the muddy murk where lights were of marginal value. Another stroke of luck was the discovery that the *Seydlitz*, whose port side was above the surface, had full coal-bunkers. The General Strike of summer 1926 had more than quadrupled the cost of coal, of which Cox needed 200 tons a week.

The first step was to apply instant cement plugs to the valves of the *Hindenburg* which had been opened to sink her; about 800 holes from porthole-size to one forty feet by twenty had to be closed. For this last, a wooden 'lid' weighing eleven tons was made ashore. It took five months to make the hull watertight; but fish ate the grease sealing the patches, and as the water began to run in, Cox's luck, for the moment, ran out. In late August, the grease having been made unpalatable to fish and the patches resealed, and after more than two weeks of pumping, the ship was flooded in a gale. They started again, but each time the ship began to rise, she listed badly. To correct this tendency two cables were attached to her mast, but they snapped with a sound like a shot from a large naval gun. They tried again. On 2 September 1926, to the cheers of the workers, she was afloat on an even keel. But a few days later another gale came and sank her again. Uncharacteristically, and £30,000 out of pocket, Cox abandoned the *Hindenburg*, if only for the time being, and turned his attention to the battlecruiser *Moltke*, the 'lucky ship' of the High Seas Fleet – which was to come within feet of bringing disaster to Scotland and proving to be the nemesis of Cox and Danks.

The *Moltke* lay upside-down in about eighty feet of water at an angle which left her level with the surface at low tide, constituting a serious danger to navigation. To get at her plates the men had to hack their way through a six-foot coating of marine vegetation. To lift her, they adopted a method which was not new but with which they were unfamiliar. The essential principle was to plug every hole in the hull and fill it with compressed air. Cox consulted Italian engineers who had successfully used this approach in raising a battleship of similar size, blown up and sunk by saboteurs in Taranto during the war. After ten days' work plugging holes and pumping, the bow began to lift, but the patches were not fully watertight, she listed and had to be allowed to sink again. Air-locks like factory chimneys, made out of old boilers joined together, were specially made at Lyness by Cox and attached to the hull so that workers could get into the hull without loss of compressed air pressure so painstakingly acquired. At the top of each tube was a hatch. The worker would close it after him and release a valve which, from below and with an ear-splitting roar, would fill the chamber in which he now found himself. When the pressure was equalised, he could open a second hatch beneath his feet and climb down the tube to the hull by means of built-in ladders. The tubes, depending on the depth of the ship to which they were attached, were up to 100 feet long and also served as decompression chambers for divers. They were used for raising all the wrecks for which the compressed air method was chosen. Pumps were lowered into the hull via the air-locks. Once a space was cleared by the creation of a pressurised air 'bubble' inside the hull, workers could go in and seal off a predetermined number of compartments, each one with its own air-lock and pumps. Working in these conditions was extraordinarily difficult. The ship, after all, was upside down, which meant that the workers were standing on the 'ceiling' and often had to erect scaffolding to get at the doors they had to seal shut. Every single aperture had to be closed. Pipes running the length of the hull were severed on either side of the bulkhead through which they passed and 'corked'. The sealing process was completed in May 1927. To counter the list, a salvaged destroyer was attached to the high side and filled with water to weigh it down. The compressed air, the rising tide and the winding-in of cables attached to the ship's gun-turrets combined to lift the *Moltke*; but the twenty cables snapped one after another. They had been worn through at the point of contact with the edge of the armoured upper deck. Metal 'cushions' were placed between the replaced cables and the edge of the deck and another attempt was made. The hulk rose slowly at first, about two feet per minute as she was sucked free of the mud at the bottom; once free of it she shot to the surface in the same way as an inflated rubber duck pops out of the bathwater when released from the bottom. Great gouts of surplus compressed air broke the surface all round the hull and the air-locks swayed wildly until the ship settled twenty feet out of the water, keel uppermost. It had taken nine months' work – no wonder that a ragged cheer broke out. In the middle of June as the ship was on tow to Lyness, she became stuck fast: one of her long 11-inch guns had snagged in the seabed. Divers went down and attached explosives to blow it away, and she was beached the next day. The workers set about stripping everything movable out of the hull to lighten the weight before the tow to the breakers. Some 3,000 tons of metal were lifted out. The Lyness Pier railway line, the only railway in the Orkneys, was extended from the pier on to the hull so that cranes could be moved into position. In May 1928 the *Moltke* was as ready for sea as she ever would be.

The unprecedented task of towing an upside-down battleship out of Scapa Flow and down the east coast of Scotland to Rosyth in the Firth of Forth, a distance of about 275 miles through some of the most treacherous water in the world, was awarded with no doubt unconscious irony to a German company, Bugsier- Reederei- und Bergungs-AG of Hamburg, which supplied three ocean-going tugs. The long air-locks had been replaced by shorter ones for the tow. Two corrugated iron huts had been built on the hull, one for the pumps to maintain the air pressure inside the hull and other gear and the other to serve as a bunkhouse for the fourteen men who were there to keep the prize afloat during the tow. Lifeboats and rafts were attached, a galley and bunks installed in the accommodation hut, and on 18 May 1928 the exhumed battleship began her last, posthumous voyage from the grave to the autopsy which was to dismember her for scrap. Inevitably, a gale blew up on the open sea and, as she wallowed helplessly in the rough water, compressed air escaped. The powerful tugs could not make headway against the fierce head-wind and the little convoy was blown backwards for a while, dragged by the leviathan of a wreck, which sank by six feet in the storm. Once it had blown over, more air was pumped into the hull and and she was raised again. The strange procession reached the Firth of Forth without further incident.

It was here that what can reasonably be described as one of the great navigational errors of modern history occurred. A civilian pilot boarded the leading tug to bring the tow safely up the Forth. Shortly afterwards, an Admiralty pilot came aboard, and a protracted jurisdictional dispute ensued. The dialogue is not recorded, but it seems the clash between the two bureaucratic prima donnas was about who was responsible for a civilian operation which was bringing a wreck to a Royal Navy dockyard for breaking up. As the words steamed up the windows on the tug's bridge, the tow-lines slackened. The *Moltke* had begun quietly to gain on her escorts, propelled by the incoming tide, which produced a current of five miles per hour. In the middle of the Firth of Forth is the rocky islet of Inchcolm, dividing the waterway in the approaches to the great Forth railway bridge. With the tow-lines slack and the impasse between the stubborn pilots rising to new heights of mutual recrimination and vituperation, nobody seemed to notice that the tug in the lead was sailing past one side of the island while the drifting hulk, behind and to one side, was floating past the other. Horrified and helpless, the tug's crew cut the tow as it snagged the island. Ahead lay the Forth Bridge, one of the great engineering triumphs of the Victorian age. Now a floating island of more than 20,000 tons of the best German steel was drifting gently but inexorably towards the central spans of the long bridge, completely out of control, still borne on the current. The petulant pilots and the trembling tugboatmen, to say nothing of the accompanying Cox, watched impotently as the inverted hulk silently bore down on the bridge. A disaster of unthinkable proportions was imminent. Cox doubtless found the time to reflect on his insurance policy taken out on the *Moltke*: the insurers had refused to cover more than two-thirds of any loss. By some miracle and much against the odds, the erstwhile 'lucky ship' drifted between the central piers of the bridge without touching them. The tugs took her in tow again and brought her to Rosyth.

Although she had failed by the narrowest of margins to destroy the Forth Bridge, the *Moltke*, as if resolved to exact payment for her desecration and the indignity of arriving upside-down, still seemed determined to be awkward. Cox had obtained permission from the Admiralty, not only to put the wreck into a naval dry dock but also to be allowed to instal her bottom-upwards, a request unique in the history of marine salvage. This was to be no easy task. At the first attempt the ship chipped the sill of the dock as she was moved out of the lock at its entrance. The damage was minor and soon made good. Now the spirit of instant and brilliant improvisation which had established itself as the hallmark of the Cox and Danks recovery operations asserted itself. A series of 'trip-wires' (made of piano-wire) were drawn taut across the approach to the dock at intervals and predetermined safe depths. They were connected to pointers on the dockside which would move if any part of the hulk's inverted upperworks made contact. Inch by inch the wreck was brought into the dock. Divers went down to stack blocks of wood in the form of pillars all along her length so that when the dock was emptied of water, the ship would come to rest on them and work could start on dismantling her. It is hardly surprising that Cox was seen literally to jump for joy. The Alloa Shipbuilding Co. took over the ship from Cox and Danks to cut her up for scrap. As the hull dried after nine years of immersion, it turned all the colours of the rainbow from the marine growths accumulated on it.

Heartened by his heart-stopping victory over the *Moltke*, Cox returned to Scapa Flow, where there was plenty of work still to be done, and plenty of setbacks to face too. In June 1928 it was the turn of the *Seydlitz* to be disinterred. The great 25,000-ton battlecruiser was as stubborn in death as she had been in life. She had taken more direct hits from the heaviest British guns in the battlecruiser fight off the Skagerrak in 1916 – nearly two dozen – than any other German ship at Jutland apart from those actually sunk. Yet she had been so splendidly built that she staggered home under her own power, her bows awash, saved by the strength of her bulkheads and the superb work of her builders, Blohm und Voss of Hamburg.

The toughest ship in the German navy now lay on her starboard side in seventy feet of water, her port flank about twenty-five feet above the surface. Her huge grey bulk often misled new visitors to Scapa Flow to mistake her for an island. Cox, pressed for money after the *Hindenburg* fiasco, decided to strip her exposed side of its 2,000 tons of armour plate and raise the *Seydlitz* sideways, selling off the first instalment of steel to raise funds. He thought the stripping would help by lowering the hulk's centre of gravity. She rose evenly after the now customary patching, sealing and application of air-locks, followed by intensive pumping. But a patch failed, and as the men working aboard jumped for their lives, her bulkheads gave way one after the other under the immense weight of inrushing water. The bow rose in the air as the *Seydlitz* went through her second death-agony; she canted over, rolled and sank upside-down amid the terrifying roar of tearing steel, crashing water and rush of escaping compressed air. The real cause of the setback was the unbalancing of her weight brought about by the stripping of her port side. Nine months' work was rendered null.

The stubborn Cox started again, determined to lift her bottom-upwards this time, like the *Moltke*. Four months later, the second attempt failed. Over the following month, forty attempts were made to raise her and they were all abandoned because she listed over dangerously as she began to rise. Unabashed by his repeated failures, Cox succumbed to his love of personal publicity and invited the press and the newsreels to come to Scape Flow on 1 November 1929 to watch the salvage of the *Seydlitz*. In the meantime, he took a holiday in Switzerland.

While he was away, his workers made a mistake. They pumped too much air into the hull – and the stubborn ship rose to the surface with all the grace of an ascending lift. Delighted, they sent a telegram to Cox at his holiday hotel to report the unexpected triumph. Cox was furious and in his answering telegram ordered that she be sunk again. Under no circumstances was he going to be deprived of his shining hour before the world's cameras. It was still a formidable gamble to take, that the lift of 1 November would work after the long record of failure, given the clearly curmudgeonly unpredictability of the scarred veteran now once again on the bottom.

Cox, a buccaneer rather than a businessman, was however prepared to risk everything for the chance to massage his ego, and, as luck would have it, the *Seydlitz* once again swept to the surface at the appointed hour, snapping ten cables as a reminder of what she could do. Nor was she defeated yet. After being gutted at Lyness, she was towed away to Rosyth and the breakers at the end of May 1930. During the last voyage, she broke away from her tow in a gale and nearly sank; then one of her sagging 11-inch guns stuck in the seabed; then she ran aground and had to be towed clear at high tide by the Hamburg tugs. All this happened before she cleared Scapa Flow. The tow took a week, twice as long as the average for the rest of the capital ships raised.

As if to console the seemingly indestructible Cox, the 24,500-ton battleship *Kaiser* came up meekly from 150 feet and, as required, at the first attempt. It was the easiest lift of all those undertaken by Cox and Danks. This time the guns in their turrets were blasted off underwater, and the ship was raised a little and allowed to drop back to drive her protruding conning tower into the hull, a method used several times on subsequent salvings. The trip to the breakers' yard passed off without incident in perfect weather. The *Bremse*, the only light cruiser to be raised, and the battlecruiser *Von der Tann* were the next to be tackled, without undue difficulty.

New dangers were dealt with; as air was pumped into the hull of a sunken ship, a highly charged vapour of oil fumes, coaldust and gas from rotting vegetation built up. Working conditions were distinctly worse than in a coalmine, and often the same danger from methane gas existed. A number of minor explosions occurred. During the work on the *Von der Tann*, three workers nearly lost their lives when their oxy-acetylene cutters set off a major blast. By the time the massive ensuing rescue operation reached them, they were up to their chins in water and breathing the last residual pocket of compressed air. For this, the men were paid 11½d an hour for a forty-eight-hour week and all found, but they loved it and respected their hard-driving boss because he worked harder than they did and was prepared to take the same risks (he often visited the wrecks in a diving suit). It is difficult now to envisage the working conditions aboard the wrecks, despite the efforts of contemporary photographers. Once the divers had introduced the pumps to create the first 'bubble' of compressed air, other workers came in to do the patching and sealing. The surfaces on which they stood were sheathed in slime; there was debris everywhere; the stench was indescribable; a lot of the work had to be done by feel because the murk was such that lights did not help when water was still present; a lot of work was done by men groping under water with only their heads above the surface; there was the danger of explosion, of collapse of bulkheads and of hidden hazards. The men who conjured the High Seas Fleet into being and who took its ships into action and later sank them were strong-minded enough; the salvage workers were made of even sterner stuff, perhaps, and the strength of character of the man who drove them was a match for any Tirpitz, Hipper or Reuter. Yet he too was to be defeated in the continuing saga of the High Seas Fleet.

At the beginning of 1930, Cox turned his attention back to the *Hindenburg*, whose superstructure must have been an irritating and constant reminder of his one major failure thus far. An examination of the hull showed that 500 or so of the 800 original patches were still good. The remaining 300 were replaced or reinforced. To stop her listing, a section of a salvaged destroyer was filled with 600 tons of concrete as a wedge under her stern. A new lift attempt caused her to heel the other way, so another wedge was sunk under the other side of the stern. On 23 July, she was brought to the surface, afloat again. The decks buckled under the strain; the stern was stove in by the huge concrete wedges, but she was successfully beached, still upright, in the now usual place in Mill Bay by Lyness. As work proceeded on gutting her of everything feasible at Lyness, Mrs McKenzie, wife of the Cox and Danks chief salvage officer, took up the pleasant if eccentric custom of installing herself in the crow's nest to knit or read a book. She gave up this habit when the jib of a crane struck the mast. In August the *Hindenburg* was towed without incident to Rosyth – the largest ship ever salvaged. Cox had spent over £30,000 to do it, and by now he had over-extended himself financially. But the removal of the great ship took away the last visible sign of the scuttling of the German High Seas Fleet from the waters of Scapa Flow, more than eleven years after it happened.

Work was proceeding on the 25,000 ton battleship *Prinzregent Luitpold* meanwhile, in great difficulty. Because her coal-bunkers had been full at the time of sinking, visibility was nil for the divers and the air introduced immediately filled with coaldust which, when dry, was as lethal as any gas. An explosion, the cause of which was never definitely established (it could have been an oxy-acetylene flame or an illicit cigarette), blew in a bulkhead and killed a carpenter. A huge rescue operation was mounted to save the other workers. As if to compensate, in July her bow broke the surface within a day of the lifting beginning. After the customary gutting, the ship was taken off to Rosyth by the German tugs in a very difficult tow, arriving on 11 May 1932. During the last voyage to the breakers, the ship persistently veered off the towing course by as much as 45 degrees, probably because she still had too much water aboard, and speed was reduced to 2½ knots.

She was the last ship ever salvaged by the formidable efforts of the sonorously named Ernest Frank Guelph Cox, the man who bought a navy. He died in 1959 at the age of seventy-six without the knighthood he had come to believe he so richly deserved. Stubborn to the end of his involvement in Scapa Flow, he gave up when his losses over investment reached about £10,000, although his business as such remained solvent. Incompetent businessman though Cox may have been, he was also undeservedly unlucky; at the time he was driven to surrender the residue of the German fleet, the price of scrap had reached a new low. Further, his equipment was superannuated and worn out. So was Cox. But he had lifted twenty-six destroyers, one light cruiser and six capital ships.

Cox's interest in the residual fleet was now taken over by his principal customer, Mr Robert McCrone, of the Alloa Shipbreaking Co., which had been buying many of the salvaged wrecks from Cox for breaking up (but, cannily, only after they had been delivered to Rosyth, whereupon the insurance risk nobly borne by Cox for the journey ceased to exist). On Cox's withdrawal, interest in the remaining ships reverted to the Admiralty, which charged McCrone a modest £1,000 for the *Bayern*, the very latest in German pre-1918 battleships with her 15-inch guns, but doubled the price for subsequent capital wrecks.

McCrone, born in 1893, shared Cox's drive and determination, obviously qualities of great value in salvaging work, but otherwise was a different kind of man altogether. He received a good technical education to college level which was complemented by sound practical experience in ship design at Vickers Armstrong and in shipyards, including dismantling scrapped warships of the Royal Navy. He won the Military Cross as a Royal Engineer officer on the Western Front. He founded his original company, Alloa Shipbreaking, with Mr Stephen Hardie, a chartered accountant, and Dr (later Sir) Donald Pollock, a surgeon-captain in the Royal Navy Volunteer Reserve, whose war service had brought him many useful Admiralty contacts. There was not enough space to work on big wrecks at Alloa, so the company took over the Rosyth Shipbreaking Co. with its facilities in the naval dockyards, which the partners added to, leasing a dry dock from the Admiralty to break up the ships delivered by Cox. The latter's withdrawal from Scapa Flow led the three men to set up Metal Industries Ltd in 1932, with McCrone as managing director and Pollock as chairman. McCrone took over the chairmanship on Pollock's departure in 1950, and himself retired in 1955. One of McCrone's chief assets was a highly developed business ability.

Metal Industries was conceived as an integrated operation which would salvage the ships, break them up and sell the scrap metal. McCrone bought out Cox's equipment at Lyness, replaced the obsolete elements in it, bought new salvage vessels, took over the workforce of 200, which was to grow to 300 at the peak of the company's activity in Scapa Flow, and reorganised the works and its methods. A large diesel engine recovered from one of the wrecks was put to work to generate power and to help manufacture liquid oxygen on the spot. In liquid form, the gas, needed for cutting, could be piped direct to the scene of operations, and the surplus was sold to other local companies. All this produced considerable savings in money and time and raised productivity to such an extent that McCrone produced a profit of £50,000 on average from each capital ship he salvaged. But he had an initial bonus too, in that just as he took over from Cox the price of scrap steel rapidly doubled. A Mr J. Robertson (no relation to the pioneer salvor at Scapa Flow) was appointed adviser on naval architecture, and a chemist called Cowan was hired to monitor the air inside the ships being worked upon for explosive methane gas.

Efficient and sophisticated though they were, especially compared with Cox's rough-and-ready methods, Metal Industries had their setbacks too. In June 1933, the *Bayern* rose to the surface prematurely after a burst compressed air pipe overfilled the hull. The shock of the sudden upward movement tore off the ship's gun-turrets, which fell to the seabed. But after eight months' work in all, on 1 September

1933, the *Bayern* made a perfect planned ascent. Once the upward force of the air inside her had sucked her free of the mud, she shot to the surface in thirty seconds. Too much air had been forced into her but no harm was done this time. The jack-in-the-box emergence of the heaviest ship in the sunken fleet (her displacement was over 28,000 tons) caused spectacular turbulence. The ship, as had become the pattern under Cox, was beached at Lyness, gutted and towed down to Rosyth. The battleship *Grosser Kurfurst* had been raised and beached in the same way just before without difficulty. The scrap from the *Bayern* fetched a total of over £110,000, nearly half of which was profit. McCrone and his men were extracting the teeth of the German fleet from the mud of Scapa Flow with a precision almost worthy of a dentist. The battleship *König Albert*, deeply embedded in cloying mud, had to be so heavily pumped with compressed air that she repeated the thirty-second ascent of the *Bayern*. Her recovery was completed without further incident, and was followed by the raising of the *Kaiserin* on 13 May 1936, another pop-up spectacular.

The next month work started on the raising of the fleet flagship, the *Friedrich der Grosse*. A diver groping his way around her hull from the inside momentarily let his imagination run away with him. He came up pale around the gills to report that he had found the ship's dungeon and that there were bones scattered among the weeds inside it. Further investigation revealed that the bones were animal and that the diver had strayed into what had been the refrigerated meat store. When the ship came up at the end of April 1937, the vessels in the Flow saluted her with their sirens and hooters.

Metal Industries meanwhile were selling their scrap from the ships to all comers on the open market. Krupp of Essen, which had made most of the armour, the guns and ships themselves, bought back some of their handiwork in the form of scrap metal and used it to help build Hitler's new navy, which drove several new battleships through the restrictions on number and tonnage imposed by the Treaty of Versailles. It is entirely possible therefore that elements of the High Seas Fleet were scuttled twice: in 1919, and again in 1939, when the battleship *Graf Spee* was sent to the bottom by her captain after the Battle of the River Plate against British cruisers. It is no less likely that the greatest of the new generation of German battleships (which were to be as frustrated as their predecessors), the *Tirpitz*, carried not only the name of the man who conceived the High Seas Fleet but also elements of its ships. At any rate this unique piece of recycling, largely the result of Pollock's good salesmanship, was one of the major applications of the recovered metal. Ironically, the Royal Navy was suffering from a shortage of armour-plate at the time.

The Nazis took a very dim view of the fact that a Hamburg towage company was still engaged in moving the hulks of German ships from their Orcadian Valhalla to the breaker's yard. Soon after they came to power, they ordered the Bugsier-, Reederei- und Bergungs-AG to withdraw its services. So in 1933 it was the turn of the Dutch, under a Captain Thomas Vet, who brought the newly completed *Zwarte Zee* (792 tons), the world's most powerful ocean-going tug at the time, and two others from Rotterdam when required. Vet took the *Friedrich der Grosse* into Rosyth after a difficult tow in August 1937. Her metal fetched over £130,000. Of the accessible big ships, that left the *Derfflinger*, the last of the battlecruisers.

She presented a special problem because she lay in deeper water than any other wreck, 150 feet down. The longest air-locks yet employed had to be made to gain access to the inverted hull, but the work, which began in July 1938, proceeded smoothly on its familiar course. To bring her up a year later from the record depth required the use of compressed air at unprecedented pressure, but she burst to the surface in *Bayern* fashion without a rupture. It was the last tribute to the skill of the Orkney salvage men, still led by McKenzie, who had transferred his allegiance to Metal Industries in 1932; it was also a remarkable attestation of the quality of the workmanship, design and materials put into her by the men of Blohm und Voss a quarter of a century earlier.

To the *Derfflinger* a peculiar world record attaches and will probably do so forever. She gained the distinction, if such it can be called, of spending more time afloat upside-down than any other ship. By the time she was raised and ready for towing away, war was once again imminent between Britain and Germany, and the Admiralty repossessed itself of its dry dock at Rosyth where all the breaking of the big ships had been done. The *Derfflinger* was therefore moored behind Rysa Island in the Flow, upside-down and kept afloat by occasional use of the compressor-pumps. She lay next to the former Grand Fleet flagship at Jutland, the *Iron Duke*, and her presence there saved the old enemy from sinking. In use as a training ship, the *Iron Duke* came under attack early in the war from three German bombers raiding the Flow, once again Britain's main naval anchorage. Severe damage was inflicted, but the indestructible McKenzie rushed his men aboard from the *Derfflinger* and patched her up so quickly that she was prevented from sinking at her moorings and beached. McKenzie was naturally recruited by the Royal Navy as a salvage officer for the duration, and he ended the war as a Commodore, RNVR, with a chestful of award ribbons.

Still afloat in 1946, the *Derfflinger*'s long deferred last voyage differed in two respects from all the others. Metal Industries mounted a complicated operation whereby she was manoeuvred into an ex-navy floating dock with a capacity of 30,000 tons, specially bought for the job of moving the old battlecruiser which had spent one war upright and another upside-down. The dock with the hulk inside it was towed, not to the Forth this time but to the Clyde on the other side of Scotland. She yielded more than 20,000 tons of scrap and proved to be the last ship of the High Seas Fleet to be raised from the bottom.

Still there to guard the graveyard of the Imperial Navy are three battleships, the *Kronprinz Wilhelm, Markgraf* and *König*, and the light cruisers *Dresden, Cöln, Karlsruhe* and *Brummer*. The three capital ships lie in a deep channel in up to 150 feet of water between Cava and the oddly-named islet Barrel of Butter, on their keels but with a marked list. The cruisers lie too deep in the mud to be worth bringing up. Given that Metal Industries decided eventually that it was not worth trying to raise the last seven, if the world were a tidier place than it is the long story of the High Seas Fleet would end at this point.

There remain, however, some intriguing loose ends to relate, even if not quite all of them can be neatly tied more than six decades after the ships went down. It is now necessary to explain how the High Seas Fleet may have extended its influence and usefulness to outer space.

The reason why recovery work has never ceased at Scapa Flow (though it has often been interrupted) is summed up in the one dread word, Hiroshima. From the time of the explosion of the first atomic bomb, the earth's atmosphere has been polluted by an unnaturally high quantity of nuclear radiation. Enormous quantities of air are sucked into the process of manufacturing steel. It follows therefore that all steel made after 1945 contains significant traces of radiation.

When it comes to the manufacture of certain delicate scientific instruments, notably devices for measuring radiation and also components for medical instruments for the treatment of cancer by controlled doses of radiation, untainted steel is an essential part of the protective shielding required to ensure maximum accuracy of readings. The world's stock of such steel, especially of the top-grade qualities used in the construction of warships and needed for shielding, is finite and dwindling. There are very few geriatric warships of pre-1945 vintage to be found throughout the world. Scapa Flow therefore remains the largest accessible source of uncontaminated steel on the planet – and not just of German origin; there still remains the British battleship *Royal Oak*, torpedoed and blown up with terrible loss of life by a German U-boat in the Flow in 1940. But she, as an official war cemetery, has been declared inviolate by the Royal Navy. That leaves the last of the German ships as the principal source.

Metal Industries closed their depot at Lyness in 1947, having disposed of the *Derfflinger* and written off the rest, and withdrew to Faslane. It was only in 1956 that they sold their interest in the last seven ships to Mr Arthur Nundy, a veteran diver who had worked for them and previously for Cox and Danks. He set up Nundy Metal Industries Ltd, which for fourteen years lived by sending down divers and skin-divers to the wrecks to break them up on the bottom. In 1970 Nundy sold out to Scapa Flow Salvage, run by Mr David Nicol and Mr Douglas Campbell; they too adopted the method of using explosives to blast off pieces from the wrecks and then hoisting them to the surface. In 1978, Scapa Flow Salvage linked up with Undersea Associates of Aberdeen, and activity was renewed with a diving survey, which confirmed that blasting had so damaged all the remaining ships that none of them could be brought up whole. The new partnership collapsed in autumn 1979 for financial reasons.

Early in 1981 the official receiver put the salvage rights up for sale and the Orkney Islands Council was urged to buy them with public funds, as the metal of the wrecks is still estimated to be worth several million pounds. It was also suggested that they could become a useful tourist attraction for the increasing number of skin-divers who go down to explore and film them. But in April 1981 Clark's Diving Services Ltd, of Lerwick in Shetland, bought four of the wrecks and leased the other three, with the intention of carrying on with the salvage work.

It is impossible to keep track of the fragments of best quality, high-density steel free of contamination which have been blasted off the seven sunken sisters and sold in many parts of the world over the past quarter of a century. As ever, some of the uses to which the metal has been put are defence secrets. It has been used to line radiation-free compartments in nuclear power stations. Wherever radioactive materials are in use, for peaceful purposes or for weapons, the level of radiation has to be monitored with extreme accuracy. And wherever that is necessary in the non-communist world, the most likely source of the vital untainted steel to project the measuring instruments is the remnant of the Imperial High Seas Fleet still in Scapa Flow. It is generally accepted in Orkney and elsewhere that some of the metal has been used in the American space programme. I asked the National Aeronautical and Space Administration to see if they could produce confirmation. They were unable to do so but were equally unable to disprove the theory. It is clear that the 'steel-mine' of Scapa Flow will retain and probably increase its value for the foreseeable future.

At least one relic of the High Seas Fleet helped the British army to escape destruction by the Germans at Dunkirk in 1940. Among the 'little ships' was a vessel with the unlikely name of *Count Dracula*. She also took part in the 40th anniversary gathering at Dunkirk in 1980, and is still going strong at the time of writing. She was once the pinnace of the *Hindenburg*.

Thus the story of the Imperial High Seas Fleet, conceived by Tirpitz, raised by the Kaiser, built by Krupp and scuttled by Reuter, remains unfinished. Its honour was salvaged by the greatest act of self-destruction in the history of navies; its ships were salvaged by the greatest recovery operation ever mounted. The last chapter in the history of the drowned ships of Scapa Flow is still not quite complete, denying this book a tidy ending. The wrecks now (2016) enjoy protection as 'scheduled monuments'. Divers are allowed to visit them but the removal of any artefact is forbidden by law.

APPENDIX I

The Ships of the Internment Formation

THE SHIPS INTERNED at Scapa Flow at the time of the order to scuttle on 21 June 1919 are listed in full below with their vital statistics. The Interned Formation (*Internierungsverband*) consisted of seventy-four vessels: five battlecruisers, all of which sank; eleven battleships, of which ten sank; eight light cruisers, of which five sank; and fifty destroyers, of which thirty-two sank. The twenty-two ships which did not go to the bottom were beached. Eventually all the battlecruisers and all the destroyers were recovered, together with eight battleships and four light cruisers. Three battleships and four light cruisers remain underwater, no longer capable of salvage except piecemeal.

THE BATTLECRUISERS

Seydlitz –

> 24,610 tons displacement, top speed 27 knots, length 606 feet, beam 94 feet, draught 27 feet; main armour 11 inches, main armament ten 11-inch guns. Built at Hamburg, launched 30 March 1912, sank 1350 GMT, salvaged November 1929.

She took twenty-two direct shell hits and one torpedo hit at Jutland but got home. She led the fleet into internment and defied more than forty attempts to raise her.

Moltke –

22,640 tons, 28 knots, 610 × 97 × 27 feet, 11-inch armour, ten 11-inch guns. Built at Hamburg, launched on 4 July 1910, sank 1310, salvaged in June 1927.

She was the first of the big ships to be raised.

Von der Tann –

19,400 tons, 28 knots, 563 × 87 × 28 feet, 9½-inch armour, eight 11-inch guns. Built at Hamburg, launched on 20 March 1909, sank 1415, salvaged in December 1930.

Derfflinger –

26,180 tons, 28 knots, 689 × 95 × 28 feet, 12-inch armour, eight 12-inch guns. Built at Hamburg, launched on 12 July 1913, sank 1445, salvaged in August 1939.

The first attempt to launch this ship failed. She holds the odd record of the longest period afloat upside-down – a good seven years. She lay the deepest of all the ships to be raised and was the last capital ship brought to the surface. She was broken up after the Second World War.

Hindenburg –

same vital statistics as the above (same class). Built at Wilhelmshaven, launched on 1 August 1915, sank 1700, salvaged in July 1930.

A fraction longer than the *Derfflinger*, this ship was the largest (but not the heaviest) in the fleet. She sank slowly on an even keel and her upperworks were part of the scenery in Scapa Flow for eleven years.

THE BATTLESHIPS

Kaiser –

24,380 tons, 21–23 knots, 564 × 95 × 27 feet, 13¾-inch armour, ten 12-inch guns. Built at Kiel, launched 22 March 1911, sank 1315, salvaged in March 1929.

She was the most orderly and shipshape large vessel in internment. She gave her name to a class, of which the next four named belonged, with almost identical statistics.

Prinzregent Luitpold –

built at Kiel, launched on 17 February 1912, sank 1330, salvaged in July 1931.

Kaiserin –

built at Kiel, launched 21 June 1909, sank 1400, salvaged in May 1936.

König Albert –

built at Danzig, launched on 27 April 1912, sank 1254, salvaged in July 1935.

Friedrich der Grosse –

built at Hamburg, launched 10 June 1911, sank 1216, salvaged in April 1937.

This riotous ship was Reuter's flagship until shortly before the scuttle, when he moved to the *Emden*. She was the first to sink.

König –

25,390 tons, 21–23 knots, 580 × 97 × 29 feet, 14-inch armour, ten 12-inch guns. Built at Wilhelmshaven, launched 1 March 1913, sank 1400.

She is still on the bottom. She gave her name to a class, of which the next three ships were also members.

Grosser Kurfürst –

built at Hamburg, launched 5 May 1913, sank 1330, salvaged in April 1933.

Kronprinz Wilhelm –

built at Kiel, launched 21 February 1914, sank 1315. Unsalvaged.

Markgraf –

built at Bremen, launched 4 June 1913, sank 1645. Unsalvaged.

Baden –

28,075 tons, 22 knots, 623 × 99 × 28 feet, 13¾-inch armour, eight 15-inch guns. Built at Danzig, launched 30 October 1915. Beached.

The *Baden* class was the mightiest type of battleship in the Imperial Navy, built as an answer to the British *Queen Elizabeth* class of super-dreadnoughts, the first to have 15-inch guns. The *Baden* joined the interned fleet only in January 1919, as a replacement for the *Mackensen* which the Allies had demanded but which was never finished.

Bayern –

Baden class. Built at Kiel, launched 18 February 1915, sank 1430, salvaged in September 1933.

These last two battleships were the heaviest in the German fleet.

THE LIGHT CRUISERS

Bremse –

4,400 tons, 28 knots, 461 × 44 × 20 feet, 1½-inch armour, four 5.9-inch guns. Built at Stettin, launched 11 March 1916, sank 1430, salvaged in November 1929.

This was the only light cruiser to be raised from the bottom. Four are still there and the other three were beached before they could sink.

Brummer –

Bremse class. Built at Stettin, launched 11 December 1915, sank 1430. Unsalvaged.

Dresden –

5,600 tons, 27–29 knots, 512 × 47 × 21 feet, 2½-inch armour, eight 5.9-inch guns. Built at Kiel, launched 25 April 1917, sank 1130. Unsalvaged.

The newest of the larger ships. *Dresden II* class.

Cöln –

Dresden II class. Built at Hamburg, launched 5 October 1916, sank 1350. Unsalvaged.

Karlsruhe –

5,440 tons, 27–28 knots, 496 × 47 × 21 feet, 2½-inch armour, eight 5.9-inch guns. Built at Wilhelmshaven, launched 31 January 1916, sank 1550. Unsalvaged.

This and the next two ships were of the *Königsberg II* class.

Nürnberg –

built at Kiel, launched 14 April 1916, drifted ashore.

Emden –

built at Bremen, launched 1 February 1916, beached. This was Reuter's flagship from which he gave the order to scuttle at 1030. She was given to France and was scrapped in 1926. She was the second light cruiser to bear the name – the first was the most successful German commerce raider of the war.

Frankfurt –

5,200 tons, 28 knots, 477 × 46 × 20 feet, 2½-inch armour, eight 5.9-inch guns. Built at Kiel, launched 20 March 1915, beached.

THE TORPEDOBOAT-DESTROYERS

(A) By flotilla

No. I Flotilla

G38, *G39*, *G40* (leader), *G86*, *V129*, *S32*

No. II Flotilla

G101, *G102*, *G103*, *V100*, *B109*, *B110* (leader), *B111*, *B112*, *G104*

No. III Flotilla

S53, *S54* (leader), *S55*, *G91*, *V70*, *V73*, *V81*, *V82*

No. VI Flotilla

No. 11 Half-Flotilla: *V43*, *V44* (leader), *V45*, *V46*, *S49*, *S50*

No. 12 Half-Flotilla: *V125*, *V126*, *V127*, *V128*, *S131*, *S132*

No. VII Flotilla

No. 13 Half-Flotilla: *S56*, *S65*, *V78*, *V83*, *G92*

No. 14 Half-Flotilla: *S136*, *S137*, *S138* (leader of flotilla and all boats), *H145*, *G89*

No. 17 Half-Flotilla

S36, *S51*, *S52*, *S60*, *V80* (half-leader)

(B) By class (with vital statistics)

Class	Members interned	Features
S31	*S32, S36*	802 tons, 33–36 knots, 261 × 27½ × 11 feet, three 3.4-inch guns, six 20-inch torpedo tubes, 24 mines
G37	*G38, G39, G40*	822 tons, 34 knots, rest as above
V43	*V43, V44, V45, V46*	852 tons, 34–36 knots, as above
S49	*S49, S50, S51, S52*	802 tons, 34–36½ knots, as above
S53	*S53, S54, S55, S56, S60, S65*	919 tons, 35–36 knots, 272½ × 27½ × 11½ feet, three 3.4-inch guns on first four, three 4.1-inch guns on last two, other details as above
V67	*V70, V73, V78, V80, V81, V82, V83*	924 tons, 34–36½ knots, 269 × 27½ × 11½ feet, three 4.1-inch guns, rest as above
G85	*G86, G89, G91, G92*	960 tons, 34 knots, 272½ × 27½ × 11½ feet, rest as above
G101	*G101, G102, G103, G104*	1,116 tons, 33½ knots, 313 × 30½ × 12 feet, as above
B109	*B109, B110, B111, B112*	1,374 tons, 36–37½ knots, 321½ × 30½ × 12½ feet, as above
V125	*V125, V126, V127, V128, V129*	924 tons, 34–34½ knots, as above
S131	*S131, S132, S136, S137, S138*	919 tons, 33–34 knots, 273 × 27½ × 12½ feet, as above
H145	*H145*	990 tons, 33½–34 knots, 277 × 27½ × 12½ feet as above

Like U-boats, German destroyers almost always bore numbers rather than names. The initial letter signifies the place of construction of the class (and all or almost all of its members).

S	means the Schickau Yard at Danzig;
G	means the Germaniawerft Yard at Kiel;
V	means the AG Vulcan Yard at Hamburg;
B	means the Blohm und Voss Yard at Hamburg;
H	means the Howaldtswerke Yard at Kiel.

The boats were moored by flotilla, chained together in pairs or threes with only one of each occupied to save fuel-oil. Their small size made them highly uncomfortable except in the calmest weather. Thirty-two sank and eighteen were beached, or settled in shallow water without going under.

APPENDIX II

The Text of the Order to Scuttle

(translated by the author)

COMMAND OF THE INTERNMENT FORMATION

Order No. P 39 17 June 1919

By officer only Personal

To: all Captains and Torpedoboat-Leader

(1) The necessary preparations for the scuttling of the ships are to be made immediately that it is ensured that the ship will sink as quickly as possible on receipt of an order.

(2) Complete preparations are to be made after the crews have been disembarked [a reference to the last reduction in crews just before the scuttle]. It will be left to the individual captain to decide to what extent individual crew-members can help and acquire knowledge of the plan without endangering these plans in any respect, e.g. by betrayal to the English.

(3) The following preparatory measures are to be adopted: all internal doors and hatches, such as gangway doors, bunker doors, doors of fore-and-aft and transverse bulkheads, hatch-companion-ways, ventilators, portholes, etc., to be kept open permanently; according to the judgement of captains, the following further steps can be taken: necessary measures to ensure the swift opening of torpedo tubes, the opening of a condenser, the opening of vents and sluices for flooding (except the latter), etc.

(4) **Boats.** If possible, no more than one boat – to take the crew – is to be prepared for untying. A (white) flag of truce is to be held ready to be taken into the boat. The boats with the disembarked crews are to make for land by the shortest route; they are to remain completely under the command of their officers.

(5) **Scuttling must take place:**

a) in the event of a forcible occupation on the part of the English;
b) on receipt of a special order from me.

(6) **On 5a.**
The captains have to consider carefully what steps are to be taken to ensure the scuttling of the ships if the English should attempt to seize the ships by surprise.
A surrender demanded by the English but not approved by me is to be regarded as a seizure by force and is to be answered with the scuttling of the ships.

(7) **On 5b.**
Orders may be given either

a) to scuttle the ship immediately, or
b) to scuttle the ship at a specified time.

(8) **Signals.** For immediate scuttling according to (7)a, the following signal will be given:

'To all Captains and Torpedoboat-Leader:
Paragraph 11. Confirm.
– Head of Internment Formation.'

This signal is to be passed on by every possible means. The scuttling of the ships must begin at once. For sinking the ship according to (7)b, the following signal will be given:

'To all Captains and Torpedoboat-Leader:
Paragraph zero nine one three. Confirm.
– Head of Internment Formation.'

In this signal the first two figures indicate the date, i.e. the 9th, and the last two figures the hour at which the scuttling should start. The time will apply from noon to noon (astronomic).

(9) **Confirmation.** Receipt of the signal is to be confirmed as follows:

'To Head of Internment Formation:
Paragraph . . . is confirmed.
– Captain, *Emden*.'

(10) As the acceptance or rejection of the Peace conditions and the reopening of hostilities is about to be decided, the next few days are specially critical. From Wednesday, 18 June, extra alertness will be necessary night and day, to observe not only every unusual action of the English but also signals from the *Emden*. As the crew cannot be fully relied upon, officers must take over the necessary lookout duties themselves.

(11) It is my intention to scuttle the ships only if the enemy should attempt to place himself in possession of them without the consent of our Government. Should our Government concur with the surrender of our ships in the peace conditions, then the ships will be given up, to the lasting shame of those who put us in this situation.

(12) Captains are to keep this document under lock and key; it must not fall into the hands of the enemy.

(signed)

v. Reuter

APPENDIX III

A Newly Discovered Eyewitness Account

In 2015 a letter from a young Royal Navy officer who witnessed the scuttling at close quarters from the British flagship HMS *Revenge* was revealed by his daughter for the first time, 96 years after the event. It will surely stand as the most vivid account on record. Sub-lieutenant Edward Hugh Markham David, then aged 18, usually called Hugh and known to his family as 'Tiny' or 'Ti' for short, wrote to his mother the day after the scuttling although, perhaps in haste, he wrongly gave the date as June 26 rather than June 22. The complete text follows, with the kind permission of his daughter, Mrs Hilary Chiswell Jones, and her family.

HMS REVENGE

Sunday 26 June 1919

My dearest Mummie,

I am writing this at sea, after witnessing perhaps the grimmest and certainly the most pathetic incident of the whole war. Out of the second-biggest and finest fleet in the world, one ship remains afloat. I will start, if you will forgive me, with a wee warning – the Admiralty have just wirelessed that they are going to publish nothing yet awhile, so if you get this account before it is published you might be tempted to publish it or something – don't on any account as it will be my exit from the service if you do.

Yesterday at 9.45 the squadron, with all destroyers at Scapa, put to sea for torpedo exercises – at 12.45 we received a wireless [message], informing us that a German battleship was sinking – we turned and at full speed dashed back to Scapa – we got back at 3.30 and the sight that met our gaze as we rounded the island of Flotta was absolutely indescribable. A good half of the German fleet had already disappeared, the water was one mass of wreckage of every description, boats, Carley floats, chairs, tables and human beings, and the *Bayern,* the largest German battleship, her bow reared vertically out of the water, was in the act of finally crashing bottomwards, which she did a few seconds later in a cloud of smoke, bursting her boilers as she went.

As soon as we appeared we were besieged by trawlers and drifters of all descriptions loaded with dead and alive Germans all piled up together – in the first a group of ragged desperados were clustered together in the bow, a little further aft sat the German Admiral von Reuter and at his feet lay a German commander stretched across a hatchway with a bullet through his head, and so on, the same in them all. I have seen men killed for the first time in my life and at that without the crash of action to keep one's spirits up, and it has made me think, God, it has made me think.

About the most dramatic moment of the whole day was the meeting of the English and German admirals. The two men were about the same height, both fine-looking and tall and as the German climbed wearily over the side there was a deadly hush on board. I was a few feet behind von Reuter so heard every word. Fremantle was of course the picture of smartness in all his admiral's trappings, whilst von Reuter, dishevelled, wet and white as a sheet, was quite the opposite – at first there was a pause, the German standing at the salute, then the following conversation:–

Fremantle: I presume you have come to surrender.
Von Reuter: I have come to surrender my men and myself

(with a sweeping gesture towards the fast sinking ships) – I have nudding [*sic*] else.

Pause.

Von Reuter: I take upon myself the whole responsibility of [*sic*] this. It is nothing to do with my officers and men – they were acting under my orders.

Fremantle: I suppose you realise that by this act of treachery (hissing voice) by this act of base treachery – you are no longer an interned enemy but my prisoner of war and as such will be treated.

Von Reuter: I understand perfectly.

Fremantle: I request you remain on the upper deck until I can dispose of you.

Von Reuter: May my flag lieutenant accompany me?

Fremantle: Yes, I grant you that.

Exit both.

All this only took a few seconds during which time I strapped a revolver round my waist, grabbed some ammunition and leapt into the drifter with an armed guard, told off to save the *Hindenburg*. The *Hindenburg* went as we were getting alongside, very nearly taking us with her. We then got alongside *Baden* who [*sic*] was going down fast, and hurried below to see what we could do to save her. We closed watertight doors which kept her up temporarily, but she eventually had to be towed ashore. We found one little German sub-lieut [enant] below who was dragged onto the upper deck, and the flag captain told him he would be shot at sunset if he did not immediately take us below and show us how to shut off the valves – his only reply was – "You can shoot me now. I do not mind." The terrible part of the whole show, to my mind, was that the Huns hadn't got a weapon between them, and it was our bounden duty to fire on them to get them back to close their valves.

You see, the ships were sunk by opening up the sea cocks at the bottom of them, and the only way we could save the ships was to force the men back on board to shut them down – none of them would go back even after half their boats' crews had been butchered – they were brave men – but we were in an awfull [*sic*] position as it was quite obvious that the Huns would die to a man rather than save their ships so that there was no point in going on firing – yet what could we do? The ships had to be saved – what the world will think I really don't know.

We are now back at Scapa having taken the prisoners down to Cromarty and turned them over to the military. A proper description is infinitely difficult to give and this one is particularly poor, but I have written it as a letter more than an accurate account of events. I will leave the many incidents to your imagination until I come home. We lost very few men; just one or two were knifed as they climbed aboard the German ships, by fanatics who had stayed behind. The whole thing has been a colossal disaster and we all await the criticism of the public on the British navy with some misgiving. I am positive that we could never have saved the ships even if we had been in harbour at the start, but the world won't believe it, I know.

I must end now – nobody knows what we are going to do but we are expecting to sail for Germany tonight to capture the fishing fleet. I have written to Dad so you need not send this on. I should like Chas to see it as I haven't got time to write him an account.

Lots of love to all
Yr [*sic*] loving son
Ti

Hugh David left the navy soon afterwards to join the RAF, created in 1918 from flying squadrons of the navy and army. He rose to the rank of Group Captain, retired from the service in 1950 and died in 1957.

A Note on Sources

THERE ARE ANY number of histories of modern Germany, and I forget how many I have read. In preparing Parts I and II of this book, I found the following works of special value.

A Short History of Germany 1815–1945 by E. J. Passant (Cambridge University Press, 1959). This has the peculiar distinction of once having been a secret document, commissioned originally by the Naval Intelligence Division of the British Admiralty and revised and expanded for general consumption.

Twentieth-Century Germany: From Bismarck to Brandt by A. J. Ryder (Macmillan, London, 1973). This is a masterly and detailed study with a comprehensive bibliography and has the added advantage of being extremely well-written.

Great Britain and the German Navy by E. L. Woodward (Oxford University Press, 1935, reprinted by Frank Cass and Co., London 1964), is a most thorough and painstaking review of the naval arms race.

From the Dreadnought to Scapa Flow: the Royal Navy in the Fisher Era, 1904–1919 by A. J. Marder (Oxford University Press, five volumes 1961–1970). The title speaks for itself – a monumental work in every sense.

The Battleship Era by Peter Padfield (Rupert Hart-Davis, London, 1972) is a very useful and technically helpful description of the development of the capital ship.

History of the First World War by B. H. Liddell Hart (Cassell,

London, 1970). The master-strategist's account is naturally mainly concerned with the struggle on land, but does not fail to put the war at sea into proper context in a book which loses no authority by reading better than most novels.

Warships and Sea Battles of World War I edited by Bernard Fitzsimons (Phoebus, London, 1973) is a collection of pertinent articles from Purnell's *History of the First World War* written by an impressive collection of experts, British and German. It contains many splendid illustrations and diagrams.

The Riddle of the Sands by Erskine Childers (Smith Elder, London, 1903). Odd to mention a spy thriller in a list like this, but this unique book combines a detailed description of the home of the Imperial Navy with a clear warning of the threat that new force presented to Britain. It helped to arouse public opinion at the time and must be one of very few works of fiction to have influenced strategy.

Jutland by Captain Donald Macintyre (Evans Brothers, London, 1957) is a much-decorated naval officer's account of the climax of the First World War at sea, complete with illustrations and maps to make it comprehensible to the landlubber.

The main section of this book, Part III, is derived principally from the priceless material in the West German Military Archive (Bundesarchiv-Militärarchiv) in Freiburg. I consulted log-books, telegram files, official reports, private accounts, German Admiralty papers and documents relating to the peace negotiations at Versailles, among many others. The best material was in a series of files with the generic title 'Scuttle of the German Fleet in Scapa Flow' covering the period 1919 to 1926. I also consulted British Admiralty records in the Public Record Office at Kew.

I also drew upon the following: *Scapa Flow: the Account of the Greatest Scuttling of all Time* by Admiral Ludwig von Reuter (Hurst and Blackett, London, 1940). This is the less than adequate translation of Reuter's recollections, originally published as *Scapa Flow:*

das Grab der deutschen Flotte (Leipzig, 1923). I relied rather more on Reuter's much lengthier report to his superiors after his return to Germany in 1920, which is in the German Archive.

Scapa Flow 1919: the end of the German Fleet by Vice-Admiral Friedrich Ruge (Ian Allan, London, 1973). Ruge was a young lieutenant on a destroyer in the interned fleet who later rose to the command of the West German Federal Navy after the Second World War. He was kind enough to assist me further in private correspondence.

There are two accounts of the salvage operation dealt with in Part IV, which I supplemented by my own researches in Orkney.

The Man who Bought a Navy by Gerald Bowman (Harrap, London, 1964) is a good account of the work of Ernest Cox.

Jutland to Junkyard by S. C. George (Patrick Stephens, Cambridge, 1973) is the work of a RAF Group Captain who explains the technical details of raising a sunken fleet with commendable clarity.

Index